Empire and Ecology in the Bengal Delta

What happens when a distant colonial power tries to tame an unfamiliar terrain in the world's largest tidal delta? This history of dramatic ecological changes in the Bengal Delta from 1760 to 1920 involves land, water and humans, tracing the stories and struggles that link them together. Pushing beyond narratives of environmental decline, Bhattacharyya argues that "property-thinking," a governing tool critical in making land and water discrete categories of bureaucratic and legal management, was at the heart of colonial urbanization and the technologies behind the draining of Calcutta. The story of ecological change is narrated alongside emergent practices of land speculation and transformation in colonial law. Bhattacharyya demonstrates how this history continues to shape our built environments with devastating consequences, as shown in the Bay of Bengal's receding coastline.

DEBJANI BHATTACHARYYA is an assistant professor of history at Drexel University. She was a junior fellow of the American Institute of India Studies and a former research fellow at the International Institute of Asian Studies, Leiden.

Studies in Environment and History

Editors

J. R. McNeill, *Georgetown University*
Edmund P. Russell, *Boston University*

Editors Emeritus

Alfred W. Crosby, *University of Texas at Austin*
Donald Worster, *University of Kansas*

Other Books in the Series

Chris Courtney *The Nature of Disaster in China: The 1931 Yangzi River Flood*
Dagomar Degroot *The Frigid Golden Age: Climate Change, the Little Ice Age, and the Dutch Republic, 1560–1720*
Edmund Russell *Greyhound Nation: A Coevolutionary History of England, 1200–1900*
Timothy J. LeCain *The Matter of History: How Things Create the Past*
Ling Zhang *The River, the Plain, and the State: An Environmental Drama in Northern Song China, 1048–1128*
Abraham H. Gibson *Feral Animals in the American South: An Evolutionary History*
Andy Bruno *The Nature of Soviet Power: An Arctic Environmental History*
David A. Bello *Across Forest, Steppe, and Mountain: Environment, Identity, and Empire in Qing China's Borderlands*
Erik Loomis *Empire of Timber: Labor Unions and the Pacific Northwest Forests*
Peter Thorsheim *Waste into Weapons: Recycling in Britain during the Second World War*
Kieko Matteson *Forests in Revolutionary France: Conservation, Community, and Conflict, 1669–1848*
Micah S. Muscolino *The Ecology of War in China: Henan Province, the Yellow River, and Beyond, 1938–1950*
George Colpitts *Pemmican Empire: Food, Trade, and the Last Bison Hunts in the North American Plains, 1780–1882*
John L. Brooke *Climate Change and the Course of Global History: A Rough Journey*
Paul Josephson et al. *An Environmental History of Russia*
Emmanuel Kreike *Environmental Infrastructure in African History: Examining the Myth of Natural Resource Management*
Gregory T. Cushman *Guano and the Opening of the Pacific World: A Global Ecological History*
Sam White *The Climate of Rebellion in the Early Modern Ottoman Empire*
Edmund Russell *Evolutionary History: Uniting History and Biology to Understand Life on Earth*
Alan Mikhail *Nature and Empire in Ottoman Egypt: An Environmental History*
Richard W. Judd *The Untilled Garden: Natural History and the Spirit of Conservation in America, 1740–1840*
James L. A. Webb, Jr. *Humanity's Burden: A Global History of Malaria*
Myrna I. Santiago *The Ecology of Oil: Environment, Labor, and the Mexican Revolution, 1900–1938*

Frank Uekoetter *The Green and the Brown: A History of Conservation in Nazi Germany*
Matthew D. Evenden *Fish versus Power: An Environmental History of the Fraser River*
Alfred W. Crosby *Ecological Imperialism: The Biological Expansion of Europe, 900–1900, second edition*
Nancy J. Jacobs *Environment, Power, and Injustice: A South African History*
Edmund Russell *War and Nature: Fighting Humans and Insects with Chemicals from World War I to Silent Spring*
Adam Rome *The Bulldozer in the Countryside: Suburban Sprawl and the Rise of American Environmentalism*
Judith Shapiro *Mao's War against Nature: Politics and the Environment in Revolutionary China*
Andrew Isenberg *The Destruction of the Bison: An Environmental History*
Thomas Dunlap *Nature and the English Diaspora*
Robert B. Marks *Tigers, Rice, Silk, and Silt: Environment and Economy in Late Imperial South China*
Mark Elvin and Tsui'jung Liu *Sediments of Time: Environment and Society in Chinese History*
Richard H. Grove *Green Imperialism: Colonial Expansion, Tropical Island Edens and the Origins of Environmentalism, 1600–1860*
Thorkild Kjærgaard *The Danish Revolution, 1500–1800: An Ecohistorical Interpretation*
Donald Worster *Nature's Economy: A History of Ecological Ideas, second edition*
Elinor G. K. Melville *A Plague of Sheep: Environmental Consequences of the Conquest of Mexico*
J. R. McNeill *The Mountains of the Mediterranean World: An Environmental History*
Theodore Steinberg *Nature Incorporated: Industrialization and the Waters of New England*
Timothy Silver *A New Face on the Countryside: Indians, Colonists, and Slaves in the South Atlantic Forests, 1500–1800*
Michael Williams *Americans and Their Forests: A Historical Geography*
Donald Worster *The Ends of the Earth: Perspectives on Modern Environmental History*
Robert Harms *Games against Nature: An Eco-Cultural History of the Nunu of Equatorial Africa*
Warren Dean *Brazil and the Struggle for Rubber: A Study in Environmental History*
Samuel P. Hays *Beauty, Health, and Permanence: Environmental Politics in the United States, 1955–1985*
Arthur F. McEvoy *The Fisherman's Problem: Ecology and Law in the California Fisheries, 1850–1980*
Kenneth F. Kiple *The Caribbean Slave: A Biological History*

Empire and Ecology in the Bengal Delta

The Making of Calcutta

DEBJANI BHATTACHARYYA
Drexel University

CAMBRIDGE
UNIVERSITY PRESS

CAMBRIDGE
UNIVERSITY PRESS

University Printing House, Cambridge CB2 8BS, United Kingdom

One Liberty Plaza, 20th Floor, New York, NY 10006, USA

477 Williamstown Road, Port Melbourne, VIC 3207, Australia

314–321, 3rd Floor, Plot 3, Splendor Forum, Jasola District Centre, New Delhi – 110025, India

79 Anson Road, #06–04/06, Singapore 079906

Cambridge University Press is part of the University of Cambridge.

It furthers the University's mission by disseminating knowledge in the pursuit of education, learning, and research at the highest international levels of excellence.

www.cambridge.org
Information on this title: www.cambridge.org/9781108425742
DOI: 10.1017/9781108348867

First published 2018

Printed in the United Kingdom by Clays, St Ives plc

A catalogue record for this publication is available from the British Library.

ISBN 978-1-108-42574-2 Hardback

Contents

Figures

Acknowledgments

This book documents the forgotten watery origins of the city of Calcutta. That would not have been possible if not for a chance encounter with Dr. Dhrubojyoti Ghosh, who asked me if the wetlands of Calcutta would ever be part of its urban history. That transformed the way I looked at the city and its vast archive. Reading Anuradha Mathur and Dilip da Cunha's work taught me to move beyond the cartographic visual literacy and listen to stories. I am beholden to their work for some of the key concepts and ideas about space I develop in the book.

This book is also the product of a large network of intellectual friendships, honest comradeship and creative kinships spread across continents. My teachers, mentors, colleagues and interlocutors offered robust criticism, coaxed the nebulous ideas into shape, brought out its creative potential and added to its historical depth. I cannot name all of them here, but they will recognize, in the pages of this book, my intellectual debt to them and my respect for the depth of their knowledge, which they allowed me to mine liberally.

The research undertaken for this book would have been inconceivable without the generous resources of the many institutions that became my home and the supportive individuals who run them. I received unstinting support from the staff and faculty at Emory University where the project was conceived and Drexel University and International Institute of Asian Studies, Leiden, where the project was completed. The American Institute of Indian Studies funded my early research for the project. Purnima Mehta, Rajender Kumar and Subir-da put at my disposal their vast network of friends and resources in Delhi and Kolkata. I am extremely grateful to archivist Sharmistha De and Bidisha Chakraborty for the battles with dust they undertook with me at the West Bengal State Archives. Ashim Mukherjee at the National Library, Kolkata, was patient with my unorthodox demands for material. Abhijit Bhattacharjee at Center for Studies in Social Sciences, Kolkata, opened up not only the Center's vast archive, but also new directions for my project. I am thankful for the resources of the Bangiya Shahitya Parishad (Kolkata), School for Cultural Texts and Records (Jadavpur University), The Asiatic Society (Kolkata), the British Library (London), Caird Library in the National Maritime Museum (London) and the National Archives of India (Delhi). I am immensely thankful to the South Asia Center at University of Pennsylvania, which offered me access to the excellent resources of the Van Pelt Library. The History Project Grant from Harvard University funded research for the final chapter, and I thank Rohit De for encouraging me to apply for it. Drexel University's College of Arts and Sciences generously supported archival research in India and the United Kingdom. Melissa Mansfield and Irene Cho were excellent in supporting and helping me in every step with procuring images for the book.

Initial ideas for this project were nurtured at Emory University where I was fortunate to work with Jeffrey Lesser, Clifton Crais and Lynne Huffer. Jeff taught me the power of simplicity in a well-crafted argument. This book would have read differently had it not been for his sharp eyes and steadfast support. Clifton pushed me to experiment methodologically and to take intellectual risks fearlessly. Lynne taught me what a non-teleological history might look like and always offered invaluable insights for structuring the argument of the book. Pamela Scully sparked my critical imagination, and her indefatigable support in critical moments shaped my intellectual trajectory and life at various stages. Joyce Fluekiger, Tom Rogers, Astrid Eckert, Elizabeth Wilson, Jonathan Prude, Gyanendra Pandey, Jamie Melton and Narayana Rao offered their insightful critique and enthusiasm for my ideas at various stages. My students at Drexel asked some of the most exciting questions about my work, and my only hope is that I have been able to answer some of them.

There are some individuals to whom I owe an immense intellectual debt. Ritu Birla has been the model of a scholar for me, and her intellectual imprint will be visible in these pages. My teachers in Kolkata, Paromita Chakravarti, Swapan Chakraborty, Supriya Chaudhuri, Abhijit Gupta and Sukanta Chaudhuri, stoked my early doubts and taught me the invaluable art of questioning. Paromitadi sustained me intellectually, creatively and sometimes financially during various phases of my career that made this book possible. Geraldine Forbes took an early interest in my project and has since offered me unwavering support. Bhavani Raman patiently discussed and constructively engaged key concepts on law and geography with me. Prathama Banjeree shared her vast knowledge of the archive and her critical insights on crafting a rich historiographical narrative. Animated exchanges, formal and informal conversations with Ishita Pande, Adam Knowles, Lena Suk, Aditya Pratap Deo, Alden Young, Rohit De, Mitra Sharafi, Faisal Chaudhry, Divya Cherian, Craig Perry, Paulomi Chakraborty and Boddhisattva Kar shaped the conceptual, creative and analytic frame of this book. Ishita Pande and Jenny Gavacs patiently read the entire draft, and Ishita's generous and incisive comments and Jenny's editorial input were invaluable to the final shape of the book.

The Center for Emotions in Max-Plank-für Bildungsforschung (Berlin) offered me an intellectually fertile space to draft the introduction and first chapter of the book while fine-tuning many of its arguments. Margrit Pernau's abundance of warmth, gentle criticism and encouragement were indispensable in the early stages of writing. More importantly, she also pushed me to start writing – no matter where I found myself. I was fortunate for the intellectual camaraderie I found there and am deeply indebted to the fellows and staff of MPIB. The draft would not have been completed without the generous support I received from International Institute of Asian Studies, Leiden. My gratitude to Philippe Peycam, Willem Vogelsang, Sandra Horst, Aarti Kawlra, Sandra Dehue and everyone else at IIAS knows no bounds.

Jatin Dua, Greg Goulding, Alessandro Battistini, Carola Lorea Erika, Mahmood Kooria, Nora Wildenauer, Emilia Sulek, Giaccomo Mantovan, Steffen Rimner, Brian Picco and Pronk offered incisive criticism, thought-provoking conversations, enthusiastic support, editorial help, food and intoxicating discussions in IIAS and Bonte Ku. I am grateful to have been surrounded by their inquiring minds during the final stages of writing in Leiden. In Nira Wickramasinghe and Sanjukta Saunderason I found invaluable intellectual friendship. At Drexel I found amazingly supportive colleagues who created a fertile environment for the development of my ideas. I thank Alden Young, Amy Slaton, Scott Knowles, Lloyd Ackert, Linda Kim, Tiago Saraiva, Gabriel Rocha, Jonathan Seitz, Don Stevens, Kathy Steen and Jonson Miller for their enthusiasm about my work. Alden has shaped many ideas in the book and read and offered original thoughts on many chapters.

I am exceptionally lucky for the support and love of old friends who offered me home, warmth, charm and endless debates: Rituparna

Sengupta, Payoshni Mitra and Trina Nileena Banerjee, Sara Ehsan, Nandita Roy and Amrita Howladar and Nagihan Haliloglu. Answesha Ghosh, Rohit De, Atreyee Majumdar, Andy Liu and Ben Siegel have discussed various stages of the project and left their individual imprints on the book. Trina, Payoshni, Aditya and Ritu have welcomed me back to their homes during my stays in Delhi and London. In Rahul Mukherjee, Beatrice Jagueri and Linda Kim I found the best writing group one can imagine. Manjita Mukharji cheered me on and indulged my inquiries about writing and academia. I am grateful for Projit Mukharji's mentorship, friendship and wit. Lynn Lees and Seth Koven invited me to be part their Delaware Valley British Studies Circle and I have found intellectual sustenance in it. I am grateful for the friendship of Sanal P. Mohan, Leilah Veviana, Deepta Shatish, Neelanjan Sircar, Navyug Gill, Casey Primel, Shatam Ray, Emma Mayer, Jill Rosenthal, Kara Moskovitz, Moyukh Chatterjee, Durba Mitra, Sunandan Kizhakke, Ajit Chittambalam, Meera Ashar, Lisa Bjorkman, Guidrex Masse, Glen Goodman and Ariel Svarch.

Infinite thanks to my editor, Lucy Rhymer and the entire team at Cambridge University Press, for their vision, guidance and patience. Lucy offered invaluable suggestions for framing and visualizing the manuscript. Series editor Edmund Russell pushed me to situate my material in a wider canvas during the early stages of writing, for which I remain extremely grateful. The two anonymous readers contributed significantly to giving the manuscript its final shape with their incisive, encouraging and helpful comments. Debdutto Banerjee was most generous for allowing me to use his image for the cover. Rudy Leon was the best indexer and I am grateful to Nichelle Rohrbacher for her diligence and care in helping me with the proofreading.

Rangan Chakraborty, in his very unassuming way, shared his thoughts, books and resources with me for more than a decade. A very special note of thanks goes to P. T. Nair and Gautam Bhadra for sharing their knowledge of Kolkata and its archives with me. Keya Dasgupta, Swapan Chakraborty, Supriya Chaudhuri, Anirban Das, William Glover, Robert Travers, Durba Ghosh, Swati Chattopadhyay, Howard Spodek, Rochona Majumdar, Sumit Guha, Indrani Chatterjee, Steven Legg, Andrew Sartori, Kapil Raj, Farina Mir, Dilip da Cunha, Anuradha Mathur, Lisa Mitchell, Surpiya-di and Sukanta-da advised and offered their suggestions at various stages of this project. Rohan D'Souza has offered comments and encouragement over many emails. I benefited from the feedback I received from AIIS's dissertation to book workshop. I am thankful to Gerri, Anand Yang, Joyce, Ramnarayan Rawat and the fellow participants of the workshop.

I presented parts of the book at various workshops and presentations, and each one of those sessions impacted small and big directions the project took. At Cornell I am grateful for the spirited engagement with my work by Raymond Craib, Chuck Geisler, Paul Nadasdy and the participants of the workshop on Contested Landscapes (2012). At Yale, David Grewal's discussion of Chapter 5 helped me place the small history of rent speculation in Calcutta as part of a larger global narrative. I also benefited from the many suggestions from the participants of the workshop on Credit, Institutions and State (2014). Seth Koven alerted me to the fluidity of both the landscape and the economy I was dealing with. Gyan Prakash encouraged me to remain aware of the sacral relations that inhere between people and places. I am grateful for the comments I received from the participants of the Nineteenth Century Workshop (2014) at Rutgers and South Asia Conference at Princeton (2016). Anupama Rao insisted I compare Bombay and Calcutta, and the final sections of the book are stronger for that. I am also thankful to the participants of the workshop on Concepts of the Urban at Columbia University (2016), Global Legal Regimes: Beyond Imperial Frames, Queen's University (2017) and Rethinking Urbanization in an Era of Climate Change at University of Pennsylvania (2017). Recent invited presentations at Presidency University and Jadavpur University, Kolkata, School for Advanced Studies in Social Sciences (EHESS) Paris, University of Humboldt, Berlin, Rotterdam University, Institute of Historical Research, London, and Cambridge University have resulted in important revisions and elaborations. An earlier version of Chapter 5, "Interwar Housing Speculation and Rent Profiteering in Colonial Calcutta," was originally published in *Comparative Studies in South Asia, Africa and the Middle East* 36, no. 3 (2016): 465–82. Copyright 2016, Duke University Press. I am thankful to Duke for allowing me to reprint the article here.

Finally, family: To each one of you I am indebted in so many ways – too many to write here. The two people who have been closest to this journey though miles apart through most of it are my mother Bandana Bhattacharyya and my father Tapan Kumar Bhattacharyya. I am so grateful for your patience with my stubbornness and impulsive wayward ways and for distracting me with many grand comforts. I dedicate this book to Baba for teaching me to take risks, Ma for giving me the courage to do so. Stan and Linda supported me with warmth and unexpected material comforts at various stages of conceiving and writing this book. With Adam, my friend and partner-in-crime, I plotted many possible directions for this book. He has read every word, pored over every sentence in the book and worried with me about the organization of my arguments – not once, but *way too many* times. He was my most ruthless, skeptical and careful reader. Thank you for adding the much-needed jocularity to this journey.

Introduction

Almanac of a Tidal Basin

> Here the perennial struggle between earth and ocean goes on, and all the secrets of land-making stand disclosed.
>
> W. W. Hunter, 1876

This is a history of forgetting in the Bengal Delta, a forgetting that enabled human design in the world's largest delta and one of its most vulnerable landscapes. From the eighteenth century on, colonial law, bureaucracy and technology eclipsed and erased the "secrets of land-making" and unmaking in Bengal's seaface, where land dissolved gradually into the vast and turbid waters of the Bay of Bengal, in order to build colonial Calcutta, the second capital of the British Empire. In the postcolonial moment, this amnesia has turned into hubris, where the aqueous and the igneous elements are now being transformed into landed concrete property. Such is the aspiration of "Atmosphere," one of the new high-rises under construction in the world's fourth-most low-lying city, Kolkata (Calcutta's current name).[1] Billed in its promotional literature as "Your Home in the Clouds," Atmosphere is selling the promise of a glistening new apartment building projecting into the sky in eastern Kolkata. The architects of Atmosphere claim that it is the first "residential sky-sculpture" in the world. Architectural design, along with India's emerging aspiration to be part of a global informational economy, manifest themselves as homogenous and smooth spaces of new satellite towns flanking Indian metropolises, often erasing their fluid ecologies.

[1] Calcutta's name was changed to Kolkata in 2001. I use Calcutta to talk about the pre-2001 city and Kolkata to refer to the current city.

In eastern Kolkata skyscrapers sprout up with unimaginable speed, tiling the landscape with uninhabited speculative properties often owned by wealthy absentee landlords. The ambition to reach for the clouds of this ethereal atmosphere is not ironic given that many owners are simply buying up air in hope of future profits. Yet this shining built landscape of glass, steel, concrete and the air waiting to turn solid is also hiding a secret beneath it. These seemingly solid structures are standing on liquid land. Even thirty years ago much of this terrain was wetlands and served as the home to fishermen and farmers. Nineteenth-century colonial officials, but also today's commuters, travelers and inhabitants, are never far from the smell of the salty bogs that flank the deltaic city of Calcutta. Yet the citizens and city's planners too often forget the delta and its salt marshes.

The Hooghly River flanks Calcutta's western banks and the city merges into vast bodies of lakes, wetlands and floodplains to its east. Calcutta is situated in a combined river system comprised of the Ganga, Brahmaputra and Meghna rivers, creating the world's largest delta. Each year these rivers carry about 40 billion cubic feet of silt on their journey to the Bay of Bengal.[2] Moreover, Calcutta is also on the edge of the world's largest mangrove estuary and rimmed by the critical wetlands of eastern India.[3] Beginning in the seventeenth century the Hooghly River appears in history as a central highway of the British Empire when Calcutta was its emporium and the capital through which commodities such as saltpeter, opium, salt, silk, cotton, jute, tea and rice moved from the east to the west. Furthermore, countless migrant laborers followed these goods in their oceanic circuits, making Calcutta a contact zone for the circulation of ideas, resources and knowledge.[4] Beyond this commercial history, the river, its water, silt and tides have shaped the city and its land market in

[2] Iftekhar Iqbal, *The Bengal Delta: Ecology, State and Social Change 1840–1943* (New York, NY: Palgrave Macmillan, 2010), 1. According to Iqbal this river system carries the highest portion of annual sediment of any world river system, amounting to about 25 percent of the world total.

[3] Sundarbans is one of the largest tidal halophytic mangrove forests in the world, about 110 miles from present Kolkata and listed as a UNESCO world heritage site in 1997. The East Kolkata Wetlands that link the city to the mangrove estuary of the Sundarbans were classified as Ramsar protected sites in 2002. However, due to entrenched governmental corruption, all the stipulations about protecting the site are flouted and the wetlands are increasingly becoming what environmentalist Dhrubajyoti Ghosh calls "real-estate-in-waiting."

[4] Kapil Raj, "The Historical Anatomy of a Contact Zone: Calcutta in the 18th Century," *Indian Economic and Social History Review* 48, no. 1 (2011): 55–82.

both implicit and explicit ways. That is the nature of deltaic cities.[5] *Empire and Ecology* conceives of the built environment of Calcutta as sedimentation of historical time, silt and human design in order to write the river and the deltaic ecology into the city's history.

Legend has it that the city was born when the ocean started churning and a tortoise, Kurma, gasped out a deep breath while being pressed by the Mandar Mountains on one side and by Ananta (the infinite) on the other. Kurma's breath spread and created the country of Kilkila, which forms the lower stretches of the Bengal Delta, where the silt-laden river moves sluggishly and the land is flat.[6] Colonial geology documents another beginning. The city, nineteenth-century earth sciences argued, was born as the eastern part of the delta started dying in the sixteenth century. Tectonic movements began silting up the Nadia rivers (consisting of Ganga, and its spill channels Jalangi, Matabhanga, Bhairab and Churni) directly north of the city, creating the vast alluvial depositions and new spaces of habitation.[7] Until the sixteenth century fishermen and falconers lived in the marshes of what is now the built-up parts of Calcutta, leaving behind a small collection of songs and couplets attesting to its long-forgotten watery origins.[8] Coincidentally, recent geological findings are more in tune with the oceanic origins of the Kurma legend and show that lower Bengal was not merely of tectonic origins but is a remnant oceanic basin.[9] Not surprisingly, locals have called this *buruniyar desh*, meaning tidal swamps or lowlands, which often remain

[5] Recent histories of Dhaka are attentive to its watery origins; see Iftekhar Iqbal, "First Master Plan for Dhaka City: An Environmental Exploration," *South Asian Chronicle* 3 (2013): 42–61; see also Sten Nilsson's discussion of Dhaka in *The New Capitals of India, Pakistan and Bangladesh* (London: Malma, 1973).

[6] No one knows the origin of the legend, which was written down by the sixteenth-century poet Kavirama in *Digbijayaprakasa*. Cited in A. K. Ray, "A Short History of Calcutta," in *Census of India 1901*, vol. 7 (Calcutta: Bengal Secretariat Press, 1902), 1.

[7] James Ferguson, "On Recent Changes in the Delta of the Ganges," *Quarterly Journal of the Geological Society* 19 (1863): 321–54. Although not a trained geologist, Ferguson wanted to develop a theory of silt deposition. Apart from Ferguson's theory, there were major debates on the behavior of Indian rivers with two separate lines of geological thinking represented by H. Leonard, Report on River Hooghly (1865) IOR/V/27/732/13 and Frederick Christian Hirst, *Report on the Nadia Rivers 1915* (Calcutta: Bengal Secretariat Press, 1916).

[8] I have explored this myth as a site for new possibilities for spatial histories of the city in Debjani Bhattacharyya, "Geography's Myth: The Many Origins of Calcutta," in *Unarchived Histories: The "Mad" and the "Trifling" in the Colonial and Postcolonial World*, ed. Gyanendra Pandey (New York, NY: Routledge, 2014), 144–58.

[9] Mahmood Alam, M. Mustafa Alam, Joseph R. Curray, M. Lutfar Rahman Chowdhury and M. Royhan Gani, "An Overview of the Sedimentary Geology of the Bengal Basin in

submerged. The ancient Indic astronomer Barahamihir termed the lower Bengal delta *samatata*, meaning land that is level with the sea.[10]

The marshes and bogs that make a patchwork of soil and swamps of Calcutta have left very few traditional records for historians. Urban marshes, floating watery soils and riverine sedimentations were not always property, but instead had to be made into property. From the eighteenth century on these soaking ecologies of land, water and biota entered the colonial archive during moments when they appeared to break down the bureaucratic logic that bounded and defined the idea of land through the legal and spatial parameters of a land grant, which conceived of land as distinct from water. Yet, as is widely known, the slow and constant movements of landmasses are not something new or unique to the area of the Bay of Bengal. Landmasses in bays and coastal areas have a different relation with water. The landmass flows, moves and challenges the fixities of cartography, ownership and territorial sovereignty.[11] However, what is specific to the Bengal Delta is the speed with which some of these geological forces have begun to unfold in the recent past.[12] As this book documents, these soaking ecologies emerged in the

Relation to the Regional Tectonic Framework and Basin-Fill History," *Sedimentary Geology* 155, no. 3 (2003): 179–208.

10 Dinesh Chandra Sen, *Studies in the Geography of Ancient and Medieval India* (Varanasi: Motilal Banarsidass, 1971), 149–58.

11 My work is deeply indebted to the radical rethinking of the relation between water, coasts and lands by Anuradha Mathur and Dilip da Cunha, *Soak: Mumbai in an Estuary* (New Delhi: Rupa Publishing, 2009) and *Mississippi Floods: Designing a Shifting Landscape* (New Haven, CT: Yale University Press, 2001). For historical works that explore the contested nature of coastline formation, see Ari Kelman, *A River and Its City: The Nature of Landscape in New Orleans* (Berkeley, CA: University of California Press, 2003); Matthew Morse Booker, *Down by the Bay: San Francisco's History between the Tides* (Berkeley, CA: University of California Press, 2013); Christopher Morris, *The Big Muddy: An Environmental History of the Mississippi and Its Peoples, from Hernando de Soto to Hurricane Katrina* (New York, NY: Oxford University Press, 2012).

12 For a literary rendering of the rivers that feed the Bengal and Bangladesh area, see Adwaita Mallabarman, *Titas Ekti Nadir Naam* (Kolkata: Punthigar, 1950). This novel was later made into an award-winning film with the same title by Ritwick Ghatak in 1973; see also Amitav Ghosh, *The Hungry Tide* (London: Harper Collins, 2004) and *The Great Derangement: Climate Change and the Unthinkable* (Chicago, IL: University of Chicago Press, 2016). While Rob Nixon's idea of "slow violence" is useful to understand the current subsidence of the Bay of Bengal coast, it nonetheless fails to capture how the mobility of the landscape has changed meanings over the course of time. Some of the movements and changes that my book captures were movements of the delta that were reinterpreted differently by different actors, with far-reaching consequences. Rob Nixon, *Slow Violence and the Environmentalism of the Poor* (Cambridge, MA: Harvard University Press, 2011).

colonial archive entangled in disputes over geographical naming, questions of ownership and economic valuation. The mobility of the landscape became central in forging laws addressing ownership in the deltaic city. The making of the property in marshes meant drying the marshes through hydraulic engineering. But simply drying did not secure tenurial or allodial rights, titles or deeds. How, then, did these temporary landscapes enter the domain of colonial urban land documentation and become part of the burgeoning property market? To answer that question, we must first investigate how property is created in marshes.

Forgetting was central to the creation of property for the extraction of value from the marshes. Technologies of fixing (legal and engineering) orchestrated a collective amnesia about the mobility of this landscape. Yet the mobility that defines the land–water relation in this region has a multifaceted history. This book will explore this history by studying how a Janus-faced legal architecture, increasing economization of space and hydraulic engineering transformed an affect-saturated and mobile delta from the latter half of the eighteenth century into a concrete space of buildings and networks of infrastructure.[13] Human intervention managed and radically transformed the slow movements of land and water in and around Calcutta since the turn of the eighteenth century. The colonial officials harnessed the mobility of the tidal delta, marked by new land formations and subsidence of lands into the river, through what I argue were specific "technologies of property." *Empire and Ecology* analyzes the historical junctures where the dramatic shifts in the river impacted the urban land market over the course of a century and a half from 1760 to 1920. It demonstrates how property law and hydraulic technologies reorganized the mobile landscape of soaking ecologies into a propertied geography through the nineteenth century in search of dried and firm land.

The dried littoral of the twentieth century was marked by economic standardization of market practices, and urban land played a significant role as property speculation catalyzed an economic crisis in colonial

[13] Scholarship on law and the colonial state, especially by Bernard Cohn and David Washbrook, has shown how colonial law was marked by two contradictions. One was the public side of the law, which sought to subordinate the rule of "Indian status" to that of the British contract. The other was the personal side, which was marked by accommodating caste, religious status and familial practices within the codified space of law. See Bernard S. Cohn, "From Indian Status to British Contract," *Journal of Economic History* 21, no. 4 (1961): 613–28; David Washbrook, "Law, State and Agrarian Society in Colonial India," *Modern Asian Studies* 15, no. 3 (1981): 649–721.

Calcutta. Economic crisis had replaced the perceived ecological crisis of the mobile landscape of the earlier century. The emergence of a speculative market in urban land in the twentieth century was intricately linked to the nineteenth-century projects of drainage, land reclamation and the repression and forgetting of these soaking ecologies. Land speculation resulted in the emergence of urban property as fictitious capital and an entity for forms of market governance for a developmental colonial state. My book, therefore, narrates a history of how colonial law and design gradually translated a complex set of human–land–water relations into a recognizable language of fixed property in the tidal swamp that was Calcutta.

The expanding property market continues to shape the current landscape of the city, whose shrinking floodplains and disappearing wetlands are part of this long history of forgetting. Colonial geologists remarked on the unique seasonal stagnation of the river as sheets of water where there was land.[14] Today we read them as urban inundation and waterlogging. Located within a shifting terrain, the history of fortifying a moving watery landscape and transforming Bengal's specific wetland ecology into that of a fixed geography of a colonial Presidency town remains largely unexplored.[15] Therefore, in order to follow the story of land and property-making we need to pay attention to the moods of the river and the movements of the littoral.[16] By making both the presence and the twentieth-century forgetting of the river and its channels in the city central

[14] Fergusson, "Recent Changes in the Delta of the Ganges," 325.

[15] Presidency towns were administrative units of the British East India Company through the nineteenth century. Prior to that they were often known as Factory Towns. Apart from Calcutta, Bombay also met a similar fate since it consisted of wetlands and archipelagos that were fortified into an island and joined with the mainland during colonialism. For a schematic overview of land reclamation in colonial Bombay, see Mariam Dossal, *Theatre of Conflict, City of Hope: Mumbai 1660 to Present Times* (New Delhi: Oxford University Press, 2010). While scholars working on Calcutta attest to the colonial trope of describing the city as a pestilential marsh, none have so far documented the history of the marshes in the colonial towns of British India. Mohit Ray's new book explores the history of the water bodies of Kolkata, but primarily in the postcolonial era. Mohit Ray, *Five Thousand Mirrors: The Water Bodies of Kolkata* (Jadavpur: Jadavpur University Press, 2015).

[16] Recent anthropological scholarship has addressed how riverine flows might affect space-making in riverfront areas. For South Asia, see Anne Rademacher, *Reigning the River: Urban Ecologies and Political Transformation in Kathmandu* (Durham, NC: Duke University Press, 2011). Franz Krause's work on the Finnish Lapland has shown the importance of the term "fluvitory" to understand these spaces of land–water admixture. See Franz Krause, "Making Space along the Kemi River: A Fluvial Geography in Finnish Lapland," *Cultural Geographies* 24, no. 2 (2017): 279–94.

to understanding urban formation in Calcutta, we can begin to view spaces beyond the one-dimensional cartographic register, which fixes spaces in time. The chapters that follow attune us to the temporality of land in this area. They explore the scientific and lived spatial practices to document how the temporality of land in the tidal swamp was repressed in our histories of property, law, cartography and design.

British colonial practices of reclaiming land and draining the city through the nineteenth century created: pockets of dried land, wealth accumulation and sites of speculation within the urban land market. The stories and struggles of the various actors who lived in and transformed the delta, including the fishermen, colonial engineers, speculators, petitioners, lawyers, property owners, municipal officials and tenants, offer material to narrate a thick history of environmental change, transformation of urban property regime and land speculation. Many of these actors used law, cartography and infrastructure in novel ways to produce these spaces as fixed – spaces that could be measured, owned and acquired as properties and thrust into a burgeoning land market.[17] Some resisted or rebuffed these changes and continued to live in these water-soaked landscape as "amphibious autochthons," to use an antiquated term coined by one of the city's early historians.[18] Above all, water, land and river resisted the various technologies of making property that abstracted the watery spaces from the ecologies within which they were located.

My story begins by documenting how infrastructural schemes carried out by private individuals in the eighteenth century and by the various urban bodies such as the Mayor's Court, the Privy Council, the Justices of Peace, Lottery Committee and the various town-planning and municipal bodies in the nineteenth century sought to make riverine water flow without stagnating and dry and thereby fix the land. Yet these projects soon translated into acts of rationalizing landholding, creating a thriving and speculative market in urban land in the ever-expanding city of Calcutta and its hinterlands. This expansion involved hydraulics, on the one hand, and forging tenuous legal legitimacy, on the other hand. By analyzing the realities created by colonial law in articulating private

[17] "Fixed," when it is used to denote property, carries a distinct meaning that I have tried to define above. It will be instructive to see how some of the Germanic languages have the sense of mobility built into the etymology of ownership where *immobilien* means property that cannot be moved (which would translate to private property like houses, buildings, etc.) and *mobilien* is often used to speak about property that can be moved, most commonly used to denote furniture.

[18] Pradip Sinha, *Calcutta in Urban History* (Calcutta: KLM Firma, 1978), 3.

property, state domain and ownership through legal and financial instruments, *Empire and Ecology* demonstrates that the manufacturing of the city's landscape was central to mediating the relation between the market in urban land and colonial statecraft.

The history of designing this particular deltaic landscape is bookended by two incidents of speculation that significantly transformed the legal, infrastructural and economic geography of Calcutta. The first instance of speculation, which would subsequently restructure the geography of the soaking ecologies, took place in the 1760s and can best be described as a venture in exploiting the geographical possibilities of the tidal landscape. Benjamin Lacam's failed venture of building a harbor in a mobile landscape in 1767 ended in a thirty-year long legal battle stretching from the corridors of the Mayor's Court in Calcutta to the halls of the House of Commons in London. The dispute unleashed a whole host of regulatory and infrastructural measures to "fix" the landscape through drying the land and developing legal classifications of its ecological varieties, which held for the next century. These intricate legal measures developed over the nineteenth century to fortify the marshes in turn consolidated Calcutta's property market, vastly transforming the legal geography of the city by the time the British Crown took over in 1857.

By the turn of the nineteenth century the nature of property speculation had clearly shifted from exploiting the geographical fluidity of the landscape to the economic intractability of landowners. Swamps around the municipal fringes of the city emerged as limits that prevented the "natural" expansion of the city, thereby contracting the land and housing markets, while also creating pockets of speculative activity.[19] Land and housing speculation by Indian property-holders and, more importantly, by newly established governmental bodies such as the Calcutta Improvement Trust, rescripted urban politics and the social and economic values of land in significant ways. This book documents the transformations in colonial laws by focusing on the environmental consolidations and legal maneuvers against the backdrop of the shifting terrain of capital accumulation.

These large shifts in colonial power and global economic concerns intersected in small plots of urban lands on the edge of the Bengal Delta.

[19] C. H. Bompas, a member of the legislative council of Bengal in 1913 and the first president of the Calcutta Improvement Trust, linked the ecological condition to the economic geography of the city C. H. Bompas, "The Work of the Calcutta Improvement Trust," *Journal of Royal Society of Arts* 75, no. 7 (1927): 199–219.

These intersections emerge in the historical archive as concerns that morphed from the perceived geographical vagaries of the delta in the early years to the so-called economic criminality of "profiteering" property owners by the twentieth century. The chapters document the everyday enactments and displays of power under the rubric of property that transferred land from the indigenous population to the state through legal fictions and condensed forms of living with mobile landscape into fixed plots and deeds. These processes continue to bolster the developmental agendas in twenty-first-century South Asian cities, reminding us that "property is a process (of making and unmaking certain kinds of relationships): that its end point is not known."[20]

Unlike previous histories of Calcutta, which treated the terrain as an inert background to explore the colonial disciplinary power and acts of resistance and negotiation, this book uncovers a different method to read urban space-making by remaining attuned to the dynamic and shifting riverine space.[21] I foreground the temporality of the landscape by locating it within the overlapping sites of their bureaucratic and legal histories. At the same time my work remains attentive to the cosmological dimensions and the affective memories that saturate these sites, whether a harbor, dock, riverbank or the silty space between land and water. Reading these sites as dense spaces of habitation subjected to scientific and legal management helps us understand how and why colonial officials read the geography of the littoral as an intractable landscape. They tried and failed to render some of these sites administratively legible in spite of an expansive

[20] Katherine Verdery, *The Vanishing Hectare: Property and Value in Postsocialist Transylvania* (Ithaca, NY: Cornell University Press, 2003), 13.

[21] Swati Chattopadhyay, *Representing Calcutta: Modernity, Nationalism, and the Colonial Uncanny* (London: Routledge, 2005) and *Unlearning the City: Infrastructure in a New Optical Field* (Minneapolis, MN: University of Minnesota Press, 2012); Partho Datta, *Planning the City: Urbanization and Reform in Calcutta, c. 1800–c. 1940* (New Delhi: Tulika Books, 2012); Pradip Sinha, *The Urban Experience, Calcutta: Essays in Honour of Professor Nisith R. Ray* (Calcutta: Riddhi-India, 1987); *Calcutta in Urban History* (Calcutta: Firma KLM Private Ltd., 1978); P. Thankappan Nair, *Calcutta: Tercentenary Bibliography* (Calcutta: Asiatic Society, 1993); *Calcutta in the 19th Century: Company's Days* (Calcutta: Firma KLM, 1989); *A History of Calcutta's Streets* (Calcutta: Firma KLM, 1987); Bhabanicharan Bandyopadhyaya, *Kalikata Kamalalay* (Calcutta: Firma KLM, 1990); Radharaman Mitra, *Kalikata Darpan* (Calcutta: Subarnarekha, 1980). This is not an exhaustive list. For an account of fishermen's history of Calcutta redolent with a watery conception of space, see Shankar De (Dnere), *Samajik O Rajnaitik Prekhapete Jele Kaibarta: Adi Kolkatar Jelepara, Jelepara-r Swang* (Kolkata: Offbeat Publishing, 2006); also see my reading of this history in Bhattacharyya, "Geography's Myth."

bureaucratic machinery of surveying, mapping and titling these spaces through property and riverine laws. The process through which a monetized value of land as alienable property replaced a *social* value in land and water as a possession involving a complex system of patronage, gifting practices, ancestral spirits and gods was neither easy nor frictionless.

Studying the processes through which colonial lawyers, surveyors and speculators consolidated and regulated the market in urban land in Calcutta opens up rich genealogies of colonial economy, urban ecology and the environment. By presenting an overlooked archive of colonial urban and environmental history of South Asia, consisting of river surveys, engineering records and court cases, along with the more traditional archive of planning and municipal documents, this book charts three interrelated processes. First, it maps the transformation of what may be called soaking ecologies into a political economy of property from the late eighteenth century onward. This process of transformation happened through a braiding together of two registers of colonial power.[22] One was the legal register, where the ecologically variable entities of silt, marshes and bogs were literally translated into landed property through naming, classifying and arbitrating ownership.[23] The other technological register consisted of the operations of drying and draining the landscape. I argue that the technological interventions into the environment of Bengal's littoral coasts cannot be understood without the attendant legal processes. Second, I document how this act of translation unfolded through the nineteenth century as colonial officials made every effort to produce a juridico-economic notion of property out of various regimes of social, ecological, ancestral and political value systems embedded in the existing ownership practices. Finally, I conclude by demonstrating that,

[22] I use the term "braiding" to imply the patchwork and plural manner through which strands of legal thinking (including common law, codified doctrines and customs) bolstered some forms of hydraulic experiments and technological innovations. I borrow this concept of braiding from Projit Bihari Mukharji, *Doctoring Traditions: Ayurveds, Small Technologies and Braided Sciences* (Chicago, IL: University of Chicago Press, 2016), 20–27.

[23] There is always an uncertainty inherent in naming geographical features, and in Chapter 1 of this book I delve into the limits of geographical taxonomy and how it emerged within law. My thinking about the limits and possibilities of different readings of place names and geography is influenced by Paul Carter, *The Road to Botany Bay: An Essay in Spatial History* (New York, NY: Faber, 1987); *Dark Writing: Geography, Performance, Design* (Honolulu, HI: University of Hawai'i Press, 2009); Greg Denning, *Islands and Beaches: Discourse on a Silent Land Marquesas 1774–1880* (Honolulu, HI: University of Hawai'i Press, 1980).

by the turn of the twentieth century, a financialized market in urban property created new social circuits and produced intrusive modes of market governance within the city, which successfully eclipsed any memory of the soaking ecologies of the littoral space.[24]

This history is critical to investigating the practices of land development that emerged beginning in the twentieth century in Calcutta. Understanding these changes in the history of urban land helps us comprehend how, through the institution of a property regime and a land market, the colonial surveyors, engineers and lawyers folded the ecological indeterminacies into colonial law as emergency provisions to acquire land through legal, but illegitimate forms of dispossession.[25] By tracing the historical changes in the property market from the eighteenth century to the current "atmosphere" of construction, this book also adds a longer genealogy to the complex inheritance of postcolonial accumulative desires that have a direct impact in reshaping the coastlines of our deltas and bays.

ALMANAC: TOWARD A NEW READING PRACTICE

The Calcutta that I uncover is one where the river and tides shape its planning, legal and economic history. It stretches beyond the bureaucrat's vision of the city ensconced in our municipal files and cartographic reproductions. By looking at the existing maps of Calcutta and the archive of planning and municipal history, it would be difficult to fully recover how legal and technological design on the marshes rearranged the geography of the urban space and transformed its ecology.[26] The visualization of practices within the cartographic register and the way we view and bound spaces in maps, tend to govern what enters the narrative of the urban space and what remains outside. Historians writing about cities therefore overwhelmingly focus on maps, surveys and the archives of town planning.[27] However, these objects of imperial rule, as we know

[24] Ritu Birla, *Stages of Capital: Law, Culture, and Market Governance in Late Colonial India* (Durham, NC: Duke University Press, 2009), 8–10 and passim.

[25] I borrow the understanding of how illegitimacies were part of colonial law from Nasser Hussain, *The Jurisprudence of Emergency: Colonialism and the Rule of Law* (Ann Arbor, MI: University of Michigan Press, 2003).

[26] For a sensitive and illuminating cartographic history of the city from its early maps to independence from British rule in 1947, see Keya Dasgupta, *Mapping Calcutta: The Collection of Maps at the Visual Archives of the Center for Studies in Social Sciences* (Calcutta: Center for Studies in Social Sciences, 2009).

[27] A classic example for Calcutta will be Datta, *Planning the City*.

them, are not, "transparent, uncontested encapsulations of a bounded [knowable] world, but performative assertions, entries into debates, points of reference for further elaboration."[28] Following John Tresch's cue on viewing systems of knowledge, (or cartographic understandings of space in this case), as a site of meshing and splitting of the quotidian, cosmological and social hierarchies, I turn to a mode of viewing and a reading practice of space that can gather together a multitude of inhabited worlds: rational, natural and spiritual. By developing a new visual literacy, we can overcome the dichotomies of Western representation versus indigenous imagination and "difference" to understand how these dynamic, mobile ways of knowing the world were in conversation with one another, and recover a hydrological urban space.[29] Before I explain this new visual literacy, let us turn to two different colonial cartographic representations, each evoking different spatial imaginations.

Empire and Ecology forges a different idiom of urban space, one that is attentive to its marshes and the tides as central in shaping the city's history. First and at its most basic, contrary to standard cartographic representations of Calcutta as depicted in Figure I.1, the landscape under analysis in this book stretches from the Sagar Islands, at the tip of the Bay of Bengal, where the early mercantile ships of the monsoon traders from Europe were docked, along the shorelines that were solidified to connect it to Calcutta (Figure I.2). This was a mobile space of sandbars, tidal sedimentations and seasonal movements of land that colonial officials fortified through infrastructures of the built environment. These fortifications happened both within the municipal city limits and outside. What linked them together was the way they worked to produce dried urban landscapes of economic value by the twentieth century. As land was dried, the tidal delta receded from our memories and the city became increasingly cut off from its watery hinterlands.

Second, *Empire and Ecology* primarily focuses on the built environments of the harbor, the urban riverbank and the drained landscape of the city and as a site to investigate the history of Calcutta's land market. Taking this space as a unit of analysis challenges the traditional organizing rubric of urban and rural, or agrarian and nonagrarian, divisions that

28 John Tresch, "Cosmologies Materialized: History of Science and History of Ideas," in *Rethinking Modern European Intellectual History*, ed. Darrin M. McMahon and Samuel Moyn (New York, NY: Oxford University Press, 2014), 163.

29 Here I am referring to Partha Chatterjee's formulation of the idea of colonial rule of difference. Partha Chatterjee, *The Nation and Its Fragments: Colonial and Postcolonial Histories* (Princeton, NJ: Princeton University Press, 1994).

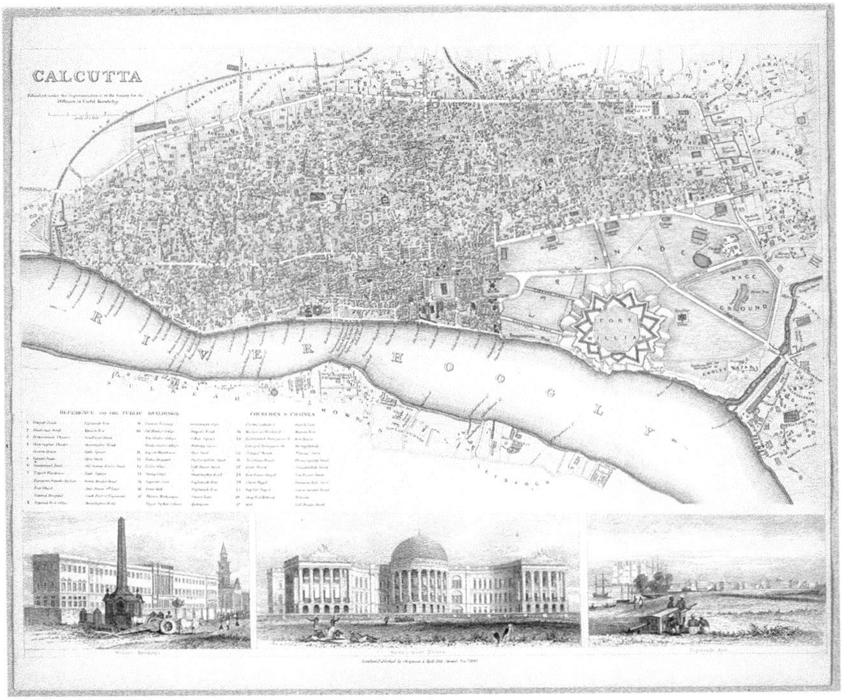

FIGURE I.1 Engraved map with pictorial inserts of Calcutta by an unknown engraver published in 1842 under the superintendence of the Society for the Diffusion of Useful Knowledge.
© British Library Board P2348.

remain central to histories of landscapes in colonial South Asia. Moreover, it helps us see Calcutta as a tidal swamp, instead of zoned spaces of residential, administrative, commercial and industrial belts that tend to dominate urban historiography, or the now-debunked black and white towns of colonial urbanism.[30] Much of the spaces discussed in this book appeared in the colonial archive with either an indeterminate ecological or economic status. According to the colonial revenue rolls and municipal documents, such spaces were either "wasteland" or economically unrealized spaces. Wastelands have a robust life in agrarian histories of South

[30] To understand how the idea of a dual city was replaced by an ideology of the sanitary city, see Ishita Pande, *Medicine, Race and Liberalism in British Bengal: Symptoms of Empire* (London: Routledge, 2010), 97–120; for a debunking of the black and white town rubric to study colonial Calcutta, see Chattopadhyay, "Limits of 'White' Town," in *Representing Calcutta*, 76–135.

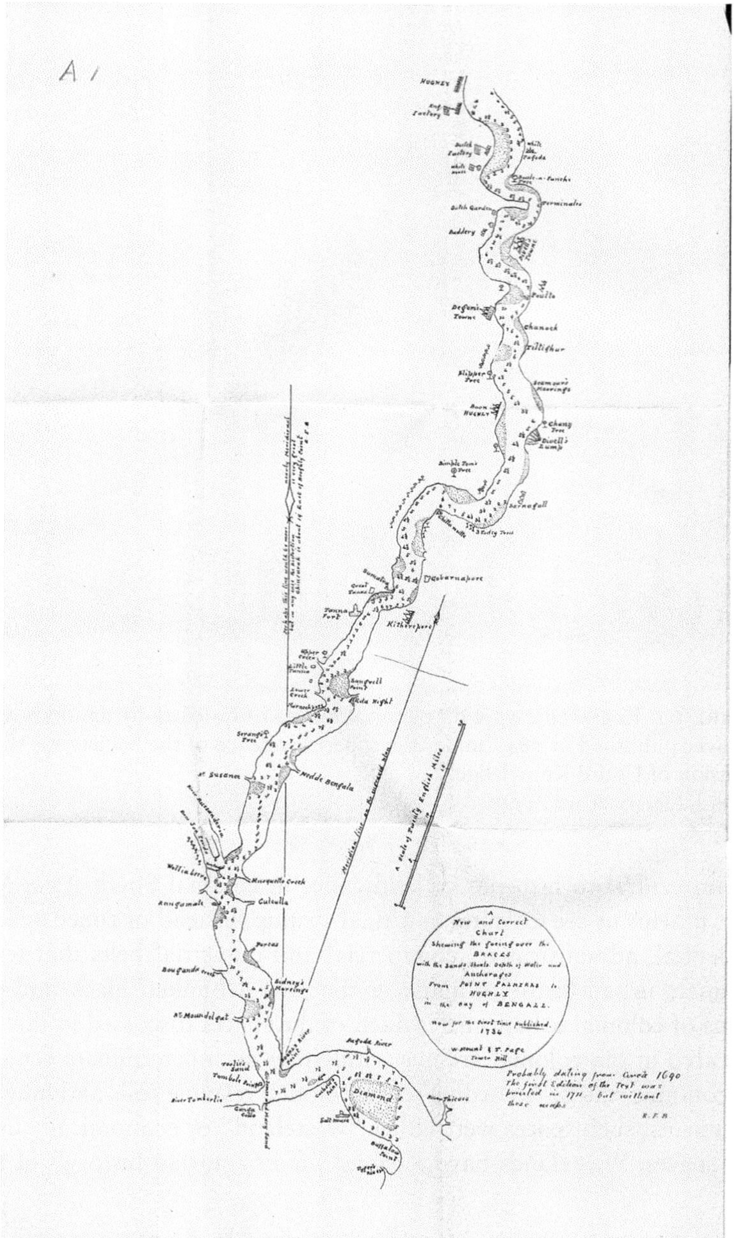

FIGURE I.2 Map of the space from Bay of Bengal to Calcutta (1790) showing the sandbars (*char*) in dotted markings. Private Papers of Barlow © British Library Board IOR/X/9128.

Asia, but those accounts fail to illuminate the fate of urban marshes and how they were produced as built environments of rails, roads and dock infrastructures that connected the delta to the city.[31] Seeing the hinterland harbors, the river and the growing city together opens up creative possibilities beyond the dominant frames to study the environment through forest/field binary and examine histories of property and land markets beyond the narrow focus of the agrarian/urban binary.[32]

Finally, what is perhaps most critical and difficult to achieve is not merely to see a wider canvas for the city that includes the hinterland, but to make the mobility of the space the anchoring point on which the interlinked marshes gather. How can history writing accomplish this attunement to the mobility of the landscape? We find clues to this in the works of landscape architects Anuradha Mathur and Dilip da Cunha who propose understanding land–water relations not as a relation of separation, but instead as that of soaking. Moreover, these thinkers force us to unlearn the cartographic-mindedness that dominates our understandings of space by developing almanac forms of reading space.[33]

[31] Vinay Gidwani, "'Waste' and the Permanent Settlement in Bengal," *Economic and Political Weekly* 27, no. 4 (1992): PE39–PE46; Judith Whitehead, "John Locke, Accumulation by Dispossession and the Governance of Colonial India," *Journal of Contemporary Asia* 42, no. 1 (2012): 1–21; Jennifer Baka, "The Political Construction of Wasteland: Governmentality, Land Acquisition and Social Inequality in South India," *Development and Change* 44, no. 2 (2013): 409–28.

[32] Ajit Menon shows that forest legislation and land settlement are rarely investigated together, thus creating the disjuncture between "field" and the "forest" within South Asian history. Ajit Menon, "Colonial Constructions of 'Agrarian Fields' and 'Forests' in the Kolli Hills," *Indian Economic and Social History Review* 41, no. 3 (2004): 315–37. Moreover, the study of urban land markets in South Asia has been strangely relegated to anthropological examination, leaving one wondering about the longer South Asian genealogies of real estate, syndicates and speculation. For some excellent analysis of the changes in urban housing and land market, see the biannual "Reviews of Urban Affairs" in *Economic and Political Weekly* from 2011 and also the works by Swagato Sarkar, "Beyond Dispossession: The Politics of Commodification of Land under Speculative Conditions," *Comparative Studies of South Asia, Africa and the Middle East* 35, no. 3 (2015): 438–50; Nikhil Anand and Anne Rademacher, "Housing in the Urban Age: Inequality and Aspiration in Mumbai," *Antipode* 43, no. 5 (2011); Llerena Guiu Searle, *Landscapes of Accumulation: Real Estate and the Neoliberal Imagination in Contemporary India* (Chicago, IL: University of Chicago Press, 2016).

[33] I would like to thank Dilip da Cunha for sharing the manuscript of his forthcoming book *Invention of the River: Alexander's Eye and Ganga's Descent* (Philadelphia, PA: University of Pennsylvania Press, 2018). For understanding land–water relations as one of soaking, I draw on Mathur and da Cunha's idea of soak; for the temporality of landscapes I have benefited from engaging with Anthony Accivatti's visualization of the river through almanacs. Mathur and da Cunha, *Soak*; Anthony Acciavatti, *Ganges Water*

As a lens into the archival material, almanacs can be a productive analytic for spatial history. Almanacs, unlike cartography, are receptive to temporality in space and its cosmological dimensions. Almanacs also endow the place–time relation with verticality toward the cosmological, while modern geographical representations freeze spaces in time and remain essentially horizontal.[34] That does not mean that almanacs are some kind of premodern, precolonial artifacts of viewing space. Rather, what I am proposing is in tune with Gautam Bhadra's seminal work on almanacs in colonial Bengal. While Bhadra's work is grounded within a textual and iconographic world of nineteenth-century almanacs, it also offers important tools for reading historical temporality not merely as the convergence of overlapping or disjointed chronologies, but also as knots where multiple possibilities exist simultaneously.[35] This pushes us beyond singular histories of urban spaces to see other inhabitations, meaning-making practices and unrealized multitudes knotted together in the movements of lands, water and the people who lived with and worked on them. Almanacs record the flux of seasons and the cosmological movements of time through space. As an analytic they operate by illuminating two lines of spatial thinking: a tidal temporality and a cosmological spatiality.

Tidal temporalities are repetitive yet unpredictable movements of water, no matter how many tidal charts we prepare. Tides also pay little respect to human-drawn boundaries between river, land, water and soil. Almanacs capture this temporality. Cartography, on the other hand, fixes moments in the movements of land and water in time and space. For instance, on a map, land and water appear as distinct entities separated by the hard lines of a coast or riverbank. For a tidal landscape like Bengal's, maps are merely a snapshot in a much longer timeline of sedimentation. Cartographic techniques fail to depict the temporality that defines land–water relations in the Bengal Delta. Applying a reticular logic to the spatial then helps recover vernacular representations of the river–water

Machine: Designing New India's Ancient River (Cambridge, MA: Harvard Design School, 2015), 4–5 and passim.

[34] Alison A. Chapman, "Marking Time: Astrology, Almanacs, and English Protestantism," *Renaissance Quarterly* 60, no. 4 (2007): 1257–90. See especially her definition of verticality, 1271.

[35] Gautam Bhadra, "Pictures in Celestial and Worldly Time: Illustrations in Nineteenth-Century Bengali Almanacs," in *New Cultural Histories of India: Materiality and Practices*, ed. Partha Chatterjee, Tapati Guha-Thakurta and Bodhisattva Kar (Delhi: Oxford University Press, 2014), 275–316. For a more in-depth treatment in Bengali, see Gautam Bhadra *Nyara Battala Jay Kaybar* (Kolkata: Chhatim Books, 2011).

relations. These relations, as delved into in Chapter 3, exceed the bounds of modern geographical understanding to reveal the limits of cartographic representations of space. Dwelling with the tidal movements make us aware of the lunar aspect, the cosmological and astrological meaning-making practices that are woven into these movements. As Bhadra and Alison Chapman show, the celestial, astrological and imperial chronologies are present in the logic of reticulation that structures the temporal order of the almanac.[36] Almanacs, attuned to the temporalities of tides, therefore hold the spatiotemporal tension between repetition and difference.

Spatially, almanacs depict a world where clock time can coexist with astrological and cosmological temporalities. This lens helps us recover a world where riverine spirits and ancestral deities splice the mundane strands of colonial bureaucracy together. Given that the Hooghly River is considered holy by the Hindu population, it is used as a site for ablutions and cremations and is dotted with small and large temple complexes. Thus, the banks of the river and its tides and the muddy space between land and water are located within multiple registers of meaning. The world this book recovers is not the familiar world where colonial rationality is opposed against an enchanted amphibious autochthony. Rather, subverting notions of colonial difference and native enchantment, I encounter cosmological worlds inhabited by colonial officials appeasing riverine spirits and Hindu deities.[37] As we see in Chapters 2 and 3, river spirits jostle with seasonal traders, colonial officials, planners and pilgrims knotting together multiple ways of inhabiting a cosmological ecology.

The historian's challenge is not to wish away the sacral and the meaning-making practices surrounding riverine activities in our secular forms of modern history writing.[38] Repeatedly, river spirits and Hindu deities make their appearance within the archive of colonial bureaucracy by making claims to spaces about to be subjected to infrastructural works. These deities must be either propitiated or accommodated in the reshaping of landscapes. This book remains open to how the cosmological appeared in human design on the landscape and how it was

[36] Bhadra, "Pictures in a Celestial and Worldly Time," 281.

[37] Gyan Prakash analyzed the mythic configurations within labor relations in colonial Bihar; see his *Bonded Histories: Genealogies of Labor and Servitude in Colonial India* (Cambridge: Cambridge University Press, 1990), esp. 34–81.

[38] Ranajit Guha, "Experience, Wonder and the Pathos of Historicality," in *History at the Limits of World History* (New York, NY: Columbia University Press, 2002), 48–74.

understood and harnessed by the colonial officials. Thus, the almanac mode of viewing spaces begins to attune us to those elusive moments when colonial bureaucracy was forced to accommodate the affective registers, including Hindu subcults and riverine spirits in their planning and legal understandings.[39] It moves us beyond the narrow readings of such sacral obstructions as the "political cunning" and calculus of resistance by the indigenous populations, and forces us to pay attention to the "non-referential forms of obligations" that inhere in these spaces.[40]

Thinking through the almanac also opens us to new ways of reading space beyond the colonized intervention and precolonial romanticism. For instance, before the advent of the Europeans, the populations living in the villages and marshes that became Calcutta, dug and widened canals for navigation. These populations moved with the fish and lived with the ebb and flow of the water. How do we understand and read those amphibious interventions into the landscape? They have left few archival records except for brief anecdotes and songs.[41] However, traces of their anthropogenic canals have been sedimented in the naming of the city, including Creek Row, a central street in Calcutta, purported to be on top of a vanished manmade canal. The same applies to Ultadanga, a Calcutta neighborhood whose name signals dry land close to the river.[42]

Empire and Ecology recovers an affect-saturated landscape that is commodified and entangled in mundane, intricate legalities. This book remains attentive to the cosmological understandings, human desires, tensions and natural spirits that shape human–land–water relations that remain outside the scope of planning documents, maps, property deeds

[39] Anthropologists have paid more attention to the intertwined world of spirits and infrastructure, while historians have read them merely as strategic resistance. For a reading of foundation sacrifice and burial narratives, see Paul G. Brewster, "The Foundation Motif in Legend, Folksong, Game and Dance," *Zeitschrift für Enthnologie* 96, no. 1 (1971): 71–89. I thank Mitra Sharafi for making me aware of this essay.

[40] Elizabeth A. Povinelli develops these two concepts in her work on aboriginal land rights in Australia. See especially chapters 1 and 2 in Elizabeth A. Povinelli, *The Cunning of Recognition: Indigenous Alterities and the Making of Australian Multiculturalism* (Durham, NC: Duke University Press, 2002).

[41] One of the few published collections of fishermen's songs, riddles and anecdotes about life in Calcutta before British arrival is in Dnere, *Samajik O Rajnaitik Prekhapete Jele Kaibarta*, 244–76.

[42] The suffix *-danga* comes from the Bengali word meaning dry land. I have explored the various connotations of place names as sites for environmental history of the city in Bhattacharyya, "Geography's Myth," 146. For a reading of the names of the various neighborhoods and the ecological past to which they attest, see also Ray, "A Short History of Calcutta," 4.

and municipal archives. Moreover, it attempts to take the challenge posed by a mobile landscape and move beyond atlas-based frameworks that define much of our modern geographical knowledge and, by default, our legal understanding of places. Such a shift questions the grounds on which our understandings of law, the market and design have solidified and forces us to confront the forms of collective amnesia that continue to bolster these understandings.

SWAMPS IN HISTORY

From the late eighteenth century on, a concerted effort to plan Calcutta was under way.[43] This effort has left us with a rich archive of colonial documents that testify how officials viewed the low-lying area of the Bengal Delta as nothing more than a pestilential marsh in need of draining. Captain Hamilton, a physician of the East India Company who toured various parts of India between the years of 1688 and 1733, commented about Calcutta that "a more unhealthful place could not be chosen on all the river."[44] Calcutta became the quintessential place of sickness and death, with the swamps regarded as the culprit. The pestilential qualities of the marshes were turned into the medicalized terminology of "miasma" when James Ranald Martin wrote his *Notes of the Medical Topography of Calcutta* in 1836. Deltaic ecology, sanitation, race and epidemiology came together under the rubric of "pathology" that needed curing and became a central idiom of colonial liberalism.[45] The "inscrutable geography" of the tidal delta entered the colonial medical archives as a disease vector when Martin proclaimed that "one would breathe thickly through the heat and miasma."[46] As the existing historiography on Calcutta argues, the medical topography of the city was deeply tied to its moral geography and its inhabitants' characters.[47] While the fear of fevers mapped out by Martin led to the intensification of

[43] Datta, *Planning the City*.

[44] Quoted in Ranald Martin, *Notes on the Medical Topography; Report of the Committee Appointed by the Right Honourable the Governor of Bengal for the Establishment of a Fever Hospital and for Inquiry into Local Management and Taxation in Calcutta* (Calcutta: Bishop College Press, 1840), 1.

[45] Pande, *Medicine, Race and Liberalism in British Bengal*, especially chapters 5 and 6.

[46] Martin, *Notes on the Medical Topography*, 18.

[47] Chattopadhyay, *Representing Calcutta*, 62–75; Datta, *Planning the City*, 127–70; Rhoads Murphey, "The City in the Swamps: Aspects of the Site and Early Growth of Calcutta," *Geographical Journal* 130, no. 2 (1964): 241–56.

draining projects, this fear also had a lasting effect in producing a particular epidemiological narrative and counternarratives about the city.[48] It is no surprise then that early historians of Calcutta employed the trope "from marshes to a metropolis" to define the British enterprise in the face of a shifting and fluid landscape.[49] More recent urban histories, which often frame their narrative around Ranald's report, depict how miasma and "olfactory imperatives" shaped imperial discourse of colonial difference and determined planning schemes that laid out the city, its neighborhoods, hinterlands and markets.[50] Still others faulted the colonial officials for their blinkered and orientalist vision for (mis)reading tropical landscapes.[51]

Epidemiology was certainly an important organizing principle, but as I argue, it was not the only organizing principle. Medico-moral frames of understanding marshes and their disappearing spaces within urban geographies are inadequate to account for the vast expenses required to dry the land. The paucity of firm land where the early British settled not only made land a precious commodity, but the cost of draining and

[48] Datta, *Planning the City*, shows the significance of Martin's medical topography in urban planning and its histories. Historians of medicine in South Asia tracked the centrality of medical topographies in the molding of race, pathology and the idea of occidental cures versus oriental therapeutics. See Pande, *Medicine, Race and Liberalism in British Bengal*, 97–147; and Arnold, *Colonizing the Body*, 11–60.

[49] Oneil Biswas, *Calcutta and Calcuttans: From Dihi to Megalopolis* (Calcutta: Firma KLM, 1992); Biren Roy, *Calcutta, 1481–1981: Marshes to Metropolis* (Calcutta: National Council of Education, 1982).

[50] I borrow "olfactory imperatives" from Alain Corbin, *Foul and Fragrant: Odor and the French Social Imagination* (Cambridge, MA: Harvard University Press, 1988), esp. 5–16. There is sometimes an overreliance on pestilence in urban historiography. See, for instance, Datta, *Planning the City*; Chattopadhyay, *Representing Calcutta*. For works on town planning and disease beyond Calcutta, see David S. Barnes, *The Great Stink of Paris and the Nineteenth-Century Struggle against Filth and Germs* (Baltimore, MD: Johns Hopkins University Press, 2006); Joseph Childers, "Foreign Matter, Imperial Filth," in *Filth: Dirt, Disgust, and Modern Life*, ed. William Cohen and Ryan Johnson (Minneapolis, MN: University of Minnesota Press, 2005); Gwendolyn Wright, *The Politics of Design in French Colonial Urbanism* (Chicago, IL: University of Chicago Press, 1991). For works that are attentive to questions of urban property and market in urban land, see Timothy Mitchell, *Rule of Experts: Egypt, Techno-Politics, Modernity* (Berkeley, CA: University of California Press, 2002); Harriet Deacon, "Racial Segregation and Medical Discourse in Nineteenth-Century Cape Town," *Journal of Southern African Studies* 22, no. 2 (1996): 287–308; Frederick Cooper, *Struggles for the City: Migrant Labour, Capital, and the State in Urban Africa* (London: Sage, 1983).

[51] Chattopadhyay, *Representing Calcutta*, 62–76.

engineering the space was not something the Company was ready to bear.[52] The ecological transformation involved in drying the city and the economic project of profit-making were joint ventures throughout the nineteenth century. Economic considerations, as much as fear of miasma, malaria, plague and cholera, directed the ideologies of draining and producing dry land. Turning the marshes into productive property was a process that in turn generated municipal profit and private capital from an expanding land market for the official and nonofficial colonists. Emphasizing the role of epidemiology alone and viewing urban planning as a response to diseases fails to account for the cornucopia of other factors: economic, ecological and legal.[53]

I trace the journey of the marshes and their materiality as they move from law courts to the engineers' desk, the land developers' snare and municipal debates. By asking how the marshes and their movement defined, challenged and reconceptualized landed property in swamps, this book is able to recover a rich legal and environmental history that went into manufacturing this space. It investigates the interplay between the conceptual legal parameters and technological interventions that began to organize and transform this space from the 1760s onward, producing a specific relation between land, water and (landed)property. Marshes significantly affected the development of a particular kind of land market in the city: a market based on reclamation and premised on a forgetting of the city's aqueous origins.

The process of land-making simultaneously became an integral process of urbanization, occurring at the technological, legal and economic registers. Let me briefly explain what I mean by land-making and its ecological salience in the lower Bengal Delta before moving on to the next section. Riverine accretion forms new land as islands in and along the banks of the river. New land formations in this tidal delta, what we have been calling marshes so far, have a specific name in this region and are called *char*.[54] Char (or *charbhumi*) and *diara* in the Gangetic plain are sedimentary or seasonal deposits, and these terms commonly refer to sandbars thrown up by the river bed, as it shifts in its course, following the annual monsoon.

52 For a short account of the speculative nature of Calcutta's land market through the nineteenth century, see Cecil H. Bompas, "The Work of the Calcutta Improvement Trust," *Journal of Royal Society of Arts* 75, no. 7 (1927): 199–219.

53 For an exception, see Pande, *Medicine, Race and Liberalism in British Bengal*, esp. 97–147 where she argues how commercial, civilizing and epidemiological interests came together in the production of a sanitary city.

54 Chapters 2 and 3 delve into how char depositions entered the urban property market.

While there is no easy translation of "char" into English, I have found it useful to understand them as "sedimentary moments" in the flows of the tidal rivers as it alerts us to the temporary and mobile nature of these spaces. Such moments manifest themselves geographically as seasonal sedimentations, only to disappear with the next annual inundation. These transient landscapes are extremely fertile, comprised of silt brought down by the Ganga River as it enters the flattest part of the plains before heading to the Bay of Bengal. Char is more commonly found in the lower Ganga Delta basin, where Calcutta is situated. Diara is prevalent in the upper delta region, and the presence of diara played a critical part in agrarian land settlements around the notorious Kosi River, a tributary of the Ganga.[55]

Within the colonial archive, char went through multiple names and legal denominations until officials finally recognized their lack of fixity as an ecological feature of the landscape. For instance, in 1825 these unstable spaces entered colonial laws of ownership as "alluvion" and "dereliction by river," as Chapters 2 and 3 of the book elaborate. Recent scholarship has noted how the marshy and transient char and diara lands were often treated either as great environmental laboratories or as wastelands during agricultural land settlement in the Bengal region.[56] These land–water admixtures, understood as "hybrid landscapes," offer us a "new metaphorical terrain," enabling alternative ways to understand human habitation and design upon the earth.[57] Unlike the agrarian char, which were subjected to very specific legal codification, the colonial officials engineered the urban char so as to erase the seasonal nature of these landscapes.[58] Built environments and infrastructure can hardly accommodate such movements in land. These seasonal lands challenged colonial cartographic registers and documentary practices of record-keeping vis-à-vis lands.

[55] There are specific geomorphological differences between the two formations, given their locations in the upper and lower reaches of the river.

[56] Christopher V. Hill, *River of Sorrow: Environment and Social Change in Riparian North India, 1770–1994* (Ann Arbor, MI: University of Michigan Press, 1997); Iqbal, *The Bengal Delta*; Kuntala Lahiri-Dutta and Gopa Samanta, *Dancing with the River: People and Life on the Chars of South Asia* (New Haven, CT: Yale University Press, 2013).

[57] Lahiri-Dutt, *Dancing with the River*, 6–11.

[58] On agrarian char, see Nitin Sinha, "Fluvial Landscape and the State: Property and the Gangetic Diaras in Colonial India, 1790s–1890s," *Environment and History* 20, no. 2 (2014): 209–37 Lahiri-Dutt, "Commodified Lands, Dangerous Water: Colonial Perceptions of Riverine Bengal," in Ursual Münster, Shiho Satsuka and Gunnel Cederlöf eds. *Asian Environments: Connections across Borders, Landscapes, and Times*. Rachel Carson Center Perspective, 2014/3). 17–23.

While the twenty-year custom is used to determine adverse possession (legal doctrine of possession in the absence of claims), char, on the other hand, could be legally considered land and documented in the land register if it had a continued existence and occupation of twelve years.[59] Very few char met this twelve-year requirement of geographical stability necessary to gain legal recognition, although people lived on and harvested these lands. Unfortunately, since they were not legally recognized as land, they left only marginal archival traces.[60] Yet the archive of urban property relations is replete with these fluvial mixes that are neither land nor water and that posed many administrative, engineering and jurisdictional challenges for the officials on ground.

Like other delta cities across the world from Hanoi to New Orleans, water and its sediments left a deep imprint in the processes of land-making in Calcutta. Studying the political economy and the legal processes of converting marshes into a propertied geography in Calcutta highlights an overlooked aspect of city formation within tidal deltas and points to new methods for understanding mobile landscapes. Moreover, the transformation of the marshes and char into property offers a unique vantage point for intervening into debates in legal and urban environmental history and its links to the colonial land market.

TECHNOLOGIES OF PROPERTY

The shifting reconfiguration of water–land–human relations connects nineteenth-century ecological changes in the Bay of Bengal region to the legal and economic history of European colonialism. How does our contemporary understanding of property law and land markets change when we try to understand them through these temporary landscapes? I view the creation of property in marshes that then reorganized social, political and economic life in the delta through what I call "technologies of property." I use the concept of property as an idea, a form to frame our thinking about land, as a language to express the division between land and water and as a legal technology to demarcate land, marsh, accretion and water. Property operated as a technology of drying, and the various

59 The legal doctrine of adverse possession is practiced in many countries and is present in both common and codified laws. For an understanding of *khas*-land (government owned), *char* land and *benami* land (land without titles; often char lands were known as *benami* land), see B. B. Mitra, ed., *The Laws of Land and Water in Bengal and Bihar, with the Case-Law Thereon* (Calcutta: Eastern Law House, 1934), 324–25.

60 Lahiri-Dutta, *Dancing with the River*, 50.

legal techniques and engineering tools ensured that property was the frame to separate land from water and to create dry land where there was silt or marsh.

The separation of land and water was both a legal confrontation beginning in the 1760s, and a technological division that took place through mapping, engineering and the production of a market in urban land over the next century and a half. In studying this separation, my book locates and analyzes the particular legal structure that made property the basis of colonial urbanization and overwhelmingly structured human relations to the land and water in the delta. While we are aware of the intellectual heritage of liberal political thought on property, I define property as a spatial device in the making of colonial power. As a heuristic term, technologies of property are therefore helpful to illuminate how the tidal landscape became readily available for economizing and organizing the space and people within a particular socioeconomic contract-based logic of the market.[61]

Framing property as a set of technologies helps us disaggregate the relationship between the act of making property and its relation to both law and space. This framing allows for the specific analysis of the spatial architecture of property. How did property become an organizing tool to govern what was perceived as indeterminate ecological spaces? Marking the marshes as property allowed for various forms of acquisition and occupation by the state while enabling new ways of staging colonial law. This was exemplified by the various laws pertaining to alluvions and regulations on land acquisition that the East India Company began to enact from the 1820s onward.[62] The political imperatives within the operation of colonial law on spaces become clear by revealing how the state calibrated its power with regard to ownership of the tidal swamp by exploiting its ecological specificity. Finally, the heuristic of viewing property as technology uncovers how this technology forged a link between

[61] My thinking about the technology and how it enframes space is indebted to Martin Heidegger, *The Question Concerning Technology, and Other Essays* (New York, NY: Harper and Row, 1977). For an application of enframing to think about markets, see Mitchell, *Rule of Experts: Egypt, Techno-Politics, Modernity*; for science, see Gyan Prakash, *Another Reason: Science and the Imagination of Modern India* (Princeton, NJ: Princeton University Press, 1999), 159–201.

[62] I have explored this in Debjani Bhattacharyya "The History of Eminent Domain in British Colonial Thought and Legal Practice in South Asia," *Economic and Political Weekly* 50, no. 50 (December 2015): 45–53.

the space and its people at particular bureaucratic sites, including surveys, petitions, litigation and speculation.

Understanding property as a particular kind of technology that enframes space in materially significant ways allows us to pose the question: Why was property conceived of as offering a fixity otherwise not available in this terrain? My book pursues this line of inquiry by exploring the following questions. Chapters 1 through 3 ask, what sort of legal arguments were available to understand and subsume ecological difference in a temporal landscape through the lens of property and state ownership? Chapter 4 investigates the conditions that made urban infrastructural spatiality in the delta a justification for creating public property in these transient spaces. Chapter 5 and the Conclusion explore the economic and legal consequences produced by the fixities of a propertied geography and drying of the delta. Broadly speaking, this book explores the diverse legal articulations that the technologies of property opened up for hydrographers, merchants, Indian landlords, town planners and speculators as they reimagined landed forms of possession over the course of the nineteenth and twentieth centuries. Pushing beyond existing understanding of colonial governmentality and subjectification, I show how administrative and legal categories developed as a response to ecological indeterminacies and took on a legal life of their own by circulating among the various actors that populate the chapters of this book.[63]

Debates over the authority, origin and rights to property have been at the heart of liberal political thought and justification of empire. From the sixteenth century on, the justification for occupation animated discussions of property, sovereignty and territoriality across the European colonies.[64]

[63] I thank the anonymous reader of the manuscript for urging me to push this line of analysis.

[64] See Richard Tuck's discussion on British political thought on the relation between property, individual rights and the power of the state. Richard Tuck, *Natural Rights Theories: Their Origin and Development* (Cambridge: Cambridge University Press, 1979), esp. 5–29. See also David Armitage, *The Ideological Origins of the British Empire* (Cambridge: Cambridge University Press, 2000); A. R. Pagden, *Lords of All the Worlds: Ideologies of Empire in Spain, Britain and France c. 1500–c. 1850* (New Haven, CT: Yale University Press, 1995). Debate on the role of the Indian colony in the development of liberal ideas in Britain has produced a vibrant scholarship. See Eric Stokes, *The English Utilitarians and India* (Oxford: Oxford University Press, 1959); Uday Mehta, *Liberalism and Empire: A Study in Nineteenth-Century British Liberal Thought* (Chicago, IL: University of Chicago Press, 1999); Thomas R. Metcalf, *Ideologies of the Raj* (Cambridge: Cambridge University Press, 1994); Karuna Mantena, *Alibis of Empire: Henry Maine and the Ends of Liberal Imperialism* (Princeton, NJ: Princeton University Press, 2010); Partha Chatterjee, *The Black Hole of Empire: History of a Global Practice of*

This discussion was primarily philosophical and based on natural law arguments offered by thinkers of the Salamanca School. With Francisco de Vitoria, the legal aspect of occupation became tied to economic questions, which were further developed in Jose Acosta's theories of progressive barbarism laying the groundwork for John Locke's labor theory of property.[65] Departing from these earlier articulations by seventeenth- and eighteenth-century thinkers of justice and sovereignty, the language of the debates around property in the nineteenth-century British Empire were decidedly historical.[66] The institution and subsequent protection of property rights was a centerpiece of the imperial "rule of law" in Britain's empire.[67] Originating in the eighteenth-century English countryside, the rule of law based on proprietary notions became a tool in the constitution of authoritative power of the state to encroach on the common rights of man.[68] While no less intrusive, property laws in colonial India became deeply enmeshed in reorganizing agrarian land holdings, credit and debt structures, and a hegemonic framework in shaping the relation between the colonial state, individuals and kinship-based communities.[69] Before

Power (Princeton, NJ: Princeton University Press, 2012); Andrew Sartori, *Liberalism in Empire: An Alternative History* (Berkeley, CA: University of California Press, 2014).

65 From the eleventh century on, medieval ecclesiastical law (specifically debates between *Glossators* and *Decretists*) attempted to delineate two aspects of occupation: what could be legally occupied at a time when European peasants were migrating to wastelands and the extent of that occupation as the poor were making claims of charity on the church. Building upon these, a natural law and juridical argument for property began to emerge with Vittorio's writings on the Iberian empires and was transformed over time through the works of Vázquez, Hugo Grotius, Samuel von Puffendorf and John Locke. For an overview, see Andrew Fitzmaurice, *Sovereignty, Property and Empire, 1500–2000* (Cambridge: Cambridge University Press, 2014), especially chapters 2–4.

66 One of the central proponents of progressive theories of property was Henry Maine's work in India. See Mantena. *Alibis of Empire*, 119–47.

67 Antony G. Hopkins, "Property Rights and Empire Building: Britain's Annexation of Lagos, 1861," *Journal of Economic History* 40, no. 4 (1980): 777–98. Ranajit Guha, *A Rule of Property for Bengal: An Essay on the Idea of Permanent Settlement* (Durham, NC: Duke University Press, 1996). Robert Travers, *Ideology and Empire in Eighteenth-Century India: The British in Bengal* (Cambridge: Cambridge University Press, 2007).

68 E. P. Thompson, *Whigs and Hunters: The Origin of the Black Act* (London: Allen Lane, 1975).

69 B. B. Chaudhuri, "Land Market in Eastern India, 1793–1940 Part II: The Changing Composition of the Landed Society," *Indian Economic & Social History Review* 12, no. 2 (1975): 133–67; Michael Mann, "A Permanent Settlement for the Ceded and Conquered Provinces: Revenue Administration in North India, 1801–1833," *Indian Economic & Social History Review* 32, no. 2 (1995): 245–69; Karuna Moy Mukerji, "Rents and Forms of Tenancy in Birbhum since the Permanent Settlement," *Indian Economic & Social History Review* 14, no. 3 (1977): 363–76; Ratna Ray, "Land Transfer and Social Change under the Permanent Settlement: A Study of Two Localities,"

the British, the Mughal Empire had developed a complex system of agrarian property holdings in establishing their right to collect revenue from agricultural land. Scholars argue while the British adapted much of the existing Mughal land tenurial system, nonetheless by rendering land an object of private property they had facilitated the transformation of existing concepts of ownership, sovereignty and value.[70] As Jon Wilson shows, regulation of the agrarian landscape was very much a search for a semantic coherence through the institution of a single type of property owner.[71] Moreover, property and inheritance laws became sites to constitute customs, in order to govern and regulate gender, family, religion and communities.[72] David Gilmartin succinctly captures the situation as follows: "Property law thus became a frame within which status-based community identities were incorporated into a structure of state-recognized individual 'right.'"[73]

Indian Economic & Social History Review 11, no. 1 (1974): 1–45; Ratnaleka Ray, "The Bengal Zamindars: Local Magnates and the State before the Permanent Settlement," *Indian Economic & Social History Review* 12, no. 3 (1975): 263–92.

70 For a discussion of the transformation of these concepts and how they impacted the nature of power and value, see Bernard S. Cohn, "Representing Authority in Victorian India," in *The Invention of Tradition*, ed. Eric Hobsbawm and Terence Ranger (Cambridge: Cambridge University Press, 1983); Robert Travers, "'The Real Value of Lands': The Nawabs, the British and Land Tax in Eighteenth-Century Bengal," *Modern Asian Studies* 38, no. 3 (2004): 517–58, and *Ideology and Empire in Eighteenth Century India: The British in Bengal* (Cambridge: Cambridge University Press, 2007).

71 Jon E. Wilson, *The Domination of Strangers: Modern Governance in Eastern India, 1780–1835* (London: Palgrave Macmillan, 2008), 104–32.

72 Rachel Sturman, *The Government of Social Life in Colonial India: Liberalism, Religious Law and Women's Rights* (Cambridge: Cambridge University Press, 2012), 35–70; Eric Lewis Beverley, "Property, Authority and Personal Law: *Waqf* in Colonial South Asia," *South Asia Research* 31 (2011): 155–82; Partha Chatterjee, *A Princely Imposter? The Strange and Universal History of the Kumar of Bhawal* (Princeton, NJ: Princeton University Press, 2002); Nicholas Dirks, "From Little King to Landlord: Property, Law and the Gift under the Madras Permanent Settlement," *Comparative Studies in Society and History* 28, no. 2 (1986): 307–33; Kasturi Malavika, "'Asceticizing' Monastic Families: Ascetic Genealogies, Property Feuds and Anglo Hindu Law in Late Colonial India," *Modern Asian Studies* 43, no. 5 (2009): 1039–83; Gregory Kozlowski, *Muslim Endowments and Society in British India* (Cambridge: Cambridge University Press, 1985) and "Muslim Women and the Control of Property in North India," in *Women and Social Reform in Modern India: A Reader*, ed. Sumit Sarkar and Tanika Sarkar (Ranikhet: Permanent Black, 2007), 20–43; Rochona Majumdar, *Marriage and Modernity: Family Values in Colonial Bengal* (Durham, NC: Duke University Press, 2009); Mytheli Sreenivas, "Conjugality and Capital: Gender, Family, and Property under Colonial Law in India," *Journal of Asia Studies* 63, no. 4 (2004): 937–60.

73 Jonathan Ocko and David Gilmartin, "State, Sovereignty, and the People: A Comparison of the 'Rule of Law' in China and India," *Journal of Asian Studies* 68, no. 1 (February 2009): 55–100, 80.

Understanding the basis of property ownership had been a protean issue for the colonial officials undertaking agrarian land settlement in Bengal from the eighteenth century on.[74] Prior to the arrival of the British in Early Modern Bengal, landed property was a site of control and wealth accumulation, and a repository of caste-based power. Landed property was encumbered by obligations and could not be bracketed simply as a contractual relation that was alienable through consent, as one could in late eighteenth-century Britain.[75] Sudipta Sen shows how in eighteenth-century Bengal, marketplaces, like property regimes, became sites of contestation between Indian and European powers. According to him, "[p]roprietary gestures in this context encompass realms of need, desire and material fulfillment that are not bound by societal rules following limited definitions of contract and narrowly construed sanguinary inheritance of property."[76] As the British realized, landed property in the Bengal countryside was as much about titles as it was about how status and caste relations and obligations were sedimented into the concept of ownership.

Such concerns about indigenous ownership practices were formulated as early as the eighteenth century. The period beginning from 1757, after the merchants of the Company defeated the Nawab (King) of Bengal, was one of unprecedented urbanization and economic standardization. Colonial officials invested enormous energy in assessing land titles for agrarian and revenue-generating land. During the agrarian revenue settlements, which began in the late eighteenth century, the problems associated with impenetrable land titles and traditions recorded in memory, customs and local usages thwarted the agents of the British East India Company in their attempts to develop a coherent and uniform revenue policy.[77] We know very little about the history of urban land, even as major

[74] For early colonial history, see Frank David Ascoli, *Early Revenue History of Bengal and the Fifth Report, 1812* (Oxford: Clarendon Press, 1917); William Bolts, *Considerations on India Affairs, Particularly Respecting the Present State of Bengal and Its Dependencies* (London: J. Almon, 1772); Harry Verelst, *View of the Rise, Progress and Present State of the English Government in Bengal, Including a Reply to the Misrepresentation of Mr. Bolts and Other Writers* (Cornhill: J. Nourse, 1772). Ranajit Guha's work alerts us to the fact that much of the early colonial history writing was an attempt to understand ownership patterns and existing property laws. Guha, *Rule of Property*, 33–36.

[75] P. J. Marshall, "Parliament and Property Rights in the Late Eighteenth-Century British Empire," in *Early Modern Conceptions of Property*, ed. John Brewer and Susan Staves (London: Routledge, 1996), 530–44.

[76] Sudipta Sen, *Empire of Free Trade: The East India Company and Making of the Colonial Marketplace* (Philadelphia, PA: University of Pennsylvania Press, 1998), 16.

[77] Guha, *A Rule of Property*, 13.

restructuring of land tenure and the forest landscape provided the basis for understanding the economic and cultural effects of colonialism in South Asia, opening up new areas of studies and offering the background for recovering the rebellious peasant, the tribal population, the *coolie* (native laborer) and plantation worker.[78]

What political and economic principles of land and property were shared across the agricultural fields, forests, plantations and ecological urban spaces? The history of land in colonial India is overwhelmingly narrated in terms of: the Permanent Settlement of 1793 in eastern India; Thomas Munro's *ryotwari* revenue settlement in southern India; the various Forest Acts; and the settlement of the itinerant population in the plantation as acts of violence and conquest.[79] These rich histories of the institution and regulation of property in the colonies have elicited robust scholarship on questions of individual rights, personal law, contract, sovereignty and territorial domination. However, beyond the major projects of conquest and extraction, what do we know about the everyday maneuvers and managerial forms of conquest that transferred the marshes into the Company coffers? Colonial power employed legal and financial instruments to create property in marshes, thus forming a market in urban land that was central to reorganizing the ecology of the space from one soaked in water to one where environmentally critical water bodies are turned into real estate on a daily basis.[80]

[78] For agrarian history, see Washbrook, "Law, State and Agrarian Society in Colonial India"; David Ludden, *Peasant History in South India* (Princeton, NJ: Princeton University Press, 1985); Bose, *Peasant Labour and Colonial Capital*; for histories of peasant rebellion, see Ranajit Guha, *Elementary Aspects of Peasant Insurgency in Colonial India* (Durham, NC: Duke University Press, 1999); Shahid Amin, *Sugarcane and Sugar in Gorakhpur: An Inquiry into Peasant Production for Capitalist Enterprise in Colonial India* (Delhi: Oxford University Press, 1984); for histories of forest settlement, see Prathama Banerjee, *Politics of Time: "Primitives" and History-Writing in a Colonial Society* (Delhi: Oxford University Press, 2006); Ajay Skaria, *Hybrid Histories: Forest, Frontiers and Wildness in Western India* (Delhi: Oxford University Press, 1998); Ramachandra Guha and Madhav Gadgil, *This Fissured Land: An Ecological History of India* (New Delhi: Oxford University Press, 1992).

[79] R. C. Dutt, "Munro and the Ryotwari Settlement in Madras 1820–27," in *The Economic History of India under Early British Rule: From the Rise of the British Power in 1757 to the Accession of Queen Victoria in 1837*, vol. 1 (London: Kegan Paul, Trench Trubner, 1902), 153–71; Ludden, *Peasant History in South India*.

[80] As I write this, road construction is planned through the East Kolkata Wetlands, violating the 2002 Ramsar ruling that deemed them protected areas. Jayanta Basu, "Roadplan Wrecks Wetlands Havoc," *Telegraph* (India), May 6, 2017, www.telegraphindia.com/1170506/jsp/calcutta/story_150080.jsp, accessed May 6, 2017.

The transformation of a tidal swamp into a thriving land market offers a broad canvas to understand how these changes left a deep imprint in the landscape and people's relation to it. An analysis of urban ecology through the framework of property is therefore important for two broad reasons. First, the rubric of property is significant for understanding the effects of modern governmental reasoning about urban space as something other than a negotiated terrain between the colonial and the colonized.[81] My work focuses on how property, as the basis of just rule and a tool against apparent environmental vagaries, emerged as a particular truth and a site for the production of law and scientific authority. Scientific and legal authority subsequently organized various kinds of indigenous practices of dwelling *with* soaking ecologies into contractual relations. Second, I contend that the regulation of the indeterminate ecologies through a rule of property is widely acknowledged but narrowly explored in urban spaces. Although the archive is replete with minor cases of battles over small patches of land between British and Indian owners, the everyday modalities of conquest remain largely overlooked in the literature. The few studies on urban property that exist focus too narrowly on the legal purview of litigation and indigenous agency in the British court to celebrate native resistance against municipal ventures. These studies have failed to locate the politics of ownership and the technologies of property as an investigation of power both within but also outside the structural logic of the state.[82] Indeed, as the first three chapters demonstrate,

[81] The rubrics of governmentality, disciplining, resistance and negotiation have been central in understanding urban space. See, for instance, Veena Talwar Oldenburg, *The Making of Colonial Lucknow, 1856–1877* (Princeton, NJ: Princeton University Press, 1984); Narayani Gupta, *Delhi between Two Empires 1803–1931: Society, Government and Urban Growth* (Delhi: Oxford University Press, 1998); Steve Legg, *Spaces of Colonialism: Delhi's Urban Governmentalities* (Malden, MA: Wiley Blackwell, 2007); Janaki Nair, *The Promise of the Metropolis: Bangalore's Twentieth Century* (New Delhi: Oxford University Press, 2005); Jyoti Hosagrahar, *Indigenous Modernities: Negotiating Architecture and Urbanism* (London: Routledge, 2005); Prashant Kidambi, *The Making of an Indian Metropolis: Colonial Governance and Public Culture in Bombay, 1890–1920* (Aldershot: Ashgate Publishing, 2007); Preeti Chopra, *A Joint Enterprise: Indian Elites and the Making of British Bombay* (Minneapolis, MN: University of Minnesota Press, 2011).

[82] See, for instance, Neils Brimnes' work on indigenous litigation in Madras and Mariam Dossal's on Bombay. Both these works have excellently revealed that the archive of Presidency towns is replete with property litigation. Neil Brimnes, "Beyond Colonial Law: Indigenous Litigation and the Contestation of Property in the Mayor's Court in Late Eighteenth-Century Madras," *Modern Asian Studies* 37, no. 3 (2003): 513–50; Mariam Dossal, *Theatre of Conflict, City of Hope*.

petitioning became a site of meaning-making about the multiple valences of ownership in these indeterminate geographies and a location to reimagine and shape political authority in the emerging urban space.

Similar to the *extra commercium* space occupied by markets in both Early Modern England and the littoral coasts of the Bay of Bengal, tenancy, landed property and ownership of marshes also straddled many registers. These registers included the economic, social, political, legal and ecological. Law had to grapple with the mobility of land, especially in urban areas and spaces of infrastructures such as docks, harbors and roads that made up the city. If, in the agrarian landscape fixing revenue became an exercise in reshaping the geography of ownership, within the urban space property itself became a technology deployed in fixing land formations and emerged as a hard line that separated the river from the city.

The history of those temporary spaces has been forgotten in the hard lines of the physical and cartographic boundaries that define our territories and the legal registers of paper documentation through which we conceive and mark our properties.[83] The history of drawing these hard lines animates the process through which land, water and all those spaces in between came to be owned in Calcutta. Thus, the hard lines on maps, titles, deeds and surveys repress the temporariness that marks the tidal space, but as this book shows, the temporariness renders propertied geography of deltas increasingly precarious. Keeping alive this temporariness and flux within the landscape challenges the permanence that bolsters our thinking around law, land market and design.

ECOLOGIES OF LAW

Property laws played a critical role in making the marshes into discrete legal elements of land and water, with each being governed by separate arenas: riparian and land laws. One of the fundamental assumptions in legal understanding is that land is fixed, firm, bounded and immobile. This assumption informs studies of imperial law and ownership practices

[83] For a fascinating study of the instability of land and its relation to property adjudication, see Theodore Steinberg, *Slide Mountain, or, the Folly of Owning Nature* (Berkeley, CA: University of California Press, 1995).

across the ideological spectrum, ranging from apologists to critics of empire.[84] Lauren Benton argued that law and geography intersected in shaping political conflicts, imperial subjecthood and communities, and it defined the scope and nature of a particular form of delegated legal authority.[85] Geography, she argues, played a central role in determining imperial sovereignty through law. Yet Renisa Mawani's work linking terrestrial and maritime terrains opens up a space to think of "aqueous and amphibian legalities" of colonialism.[86] Amphibious spaces reorganize the relation between law and geography. The contestations over the apparent ecological vagaries of the tidal delta reveal that geography, instead of determining and defining the legal architecture, became a productive site for legal argument and reasoning, as I show in Chapter 1. Indeed, law courts also became the site for the production of geographical knowledge.

Empire and Ecology demonstrates that the question of ownership in this active delta had to grapple with an elemental problem of what constitutes land in a watery terrain. Imperial legal thought was repeatedly challenged in the face of multiple ownership practices in which property could not simply be defined as a concern for law or a concern for the municipal revenue department. Even though the Company merchants had begun codifying personal Hindu and Islamic laws from the latter half of the eighteenth century, the legal regulation of urban property continued to be a vexing issue for the colonial government through the nineteenth century and thus serves as a salient vantage point for comprehending colonial law's unsettled relation to property.[87] Focusing on the mobility

[84] For recent attempts at looking at the circuits of law and legal realities beyond solid land, see Lauren A. Benton, *A Search for Sovereignty: Law and Geography in European Empires, 1400–1900* (Cambridge: Cambridge University Press, 2009); Iza Hussin, "Circulations of Law: Cosmopolitan Elites, Global Repertoires, Local Vernaculars," *Law and History Review* 32 no. 4 (2014): 773 –95; Kerry Ward, *Networks of Empire: Forced Migration in the Dutch East India Company* (Cambridge: Cambridge University Press, 2009); and the special issue by Renisa Mawani and Iza Hussin, "The Travels of Law: Indian Ocean Itineraries," *Law and History Review* 32, no. 4 (2014).

[85] Benton, *A Search for Sovereignty*, 30–38.

[86] Renisa Mawani, "Law and Settler Colonialism, and 'the Forgotten Space' of Maritime Worlds," *Annual Review of Law and Social Science* 12 (2013): 107–31.

[87] Michael R. Anderson, "Classification and Coercions: Themes in South Asian Legal Studies in the 1980s," *South Asia Research* 10, no. 2 (1990): 158–77; Michael R. Anderson and Sumit Guha, eds., *Changing Concepts of Rights and Justice in South Asia* (New Delhi: Oxford University Press, 2000); Bernard Cohn, "The Census, Social Structure and Objectification in South Asia," in Bernard Cohn, *An Anthropologist among the Historians and Other Essays* (Delhi: Oxford University Press, 1987),

of land raises important questions about the exact nature of the colonial laws pertaining to marshes.

Until the second half of the nineteenth century, colonial law, especially in relation to these marshy spaces, was not a fully formed entity, because geological knowledge had yet to offer concepts about tidal movements. In order to understand the practices of law, I turn to the enunciations, information, utterances, reasoning and understandings about these spaces as they are expressed in the court petitions and debates from that era.[88] A vast range of actors including sailors, engineers, colonial bureaucrats, hydrographers, surveyors, cartographers and the Indians living with these marshes were involved in producing knowledge about these spaces both within and outside the law courts. Rather than merely symbolizing "remoteness and wildness," the shifting sandbars of the Bengal Delta created a legal circuit between Calcutta and London, as I show in Chapter 1.[89] Geography became a site of legal evidence where petitioners claimed, debated and made sense of human–land–water relations in the area. Colonial and metropolitan lawyers then translated these sets of geographical information into specific legal doctrines governing ownership of land, water and marshes. Ultimately, the question for the British was how to occupy marshy spaces "legally," even if not "legitimately."[90]

Approaching this process of law-making in studying how the British came to occupy the marshes or "wastelands," as they called them, therefore shifts the focus away from the dominant frames of justice and the

224–54; Neeladri Bhattacharya, "Remaking Custom: The Discourse and Practice of Colonial Codification," in *Tradition, Dissent and Ideology: Essays in Honour of Romila Thapar*, ed. R. Champakalakshmi and S. Gopal (Delhi: Oxford University Press, 1996), 20–51; Nandini Bhattacharyya-Panda, *Appropriation and Invention of Tradition: The East India Company and Hindu Law in Early Colonial Bengal* (Delhi: Oxford University Press, 2008).

88 Niklas Luhmann, *Law as a Social System*, ed. Klaus A. Ziegert and Fatima Kastner (Oxford: Oxford University Press, 2004).

89 For the concepts of layered sovereignty and plural legality produced by geographical "remoteness," see Benton, *A Search for Sovereignty*, 32–34 and 292–99.

90 The contradiction between legitimate and legal in colonial law has been explored by Partha Chatterjee, who showed that the structure of colonial state was such that it could not complete its own project of modernizing colonial society through a rule of law. Law was universal, yet it had to address colonial difference. This has been further expanded by Nasser Hussain, who reads the site of difference within law as a site to embed emergency provisions or what he has termed the "plenitude of prerogative." Partha Chatterjee, *The Nation and Its Fragments: Colonial and Postcolonial Histories* (Princeton, NJ: Princeton University Press, 1993), 14–19; Hussain, *The Jurisprudence of Emergency*, 26–30.

doctrine of *terra nullius* that have prevailed in discussions on geography's relation to law. Recent debates have debunked the concept of *terra nullius* and shown it to be an "invented legal event" that obfuscates the myriad processes of juridical acquisition and colonial dispossession.[91] It is well known that the doctrine of *terra nullius*, or *vaccum domicilum* regarding British imperial expansion in North America, Africa and Australia, proceeded along a different line from the debates in South Asian history. In the South Asian context, the question of justice did not rest so much on ownerless lands and the possibilities of occupation, but instead revolved around the nature of ownership, ancient rights, customary practices and the introduction of a contract-based documentary regime backing the idea of ownership. Historians have argued that, in the process of ushering in the modern propertied citizen in South Asia, colonial officials translated a whole host of customs into contracts, resulting in land alienation.[92] Yet most of these rich debates around property and territoriality fail to address if and how geographical specificity inflected the discussion around law and imperial acquisition. It is surprising that the question regarding the materiality of land has largely been ignored in the intellectual debates about property in South Asia. This is especially astonishing given that what the Company designated as "wasteland," i.e., marshes, jungles, high grass and reeds, as well as char and diara, often became another term of unsettled land, or *Khasmahal* (land under direct government management) in the agrarian revenue rolls.

Early merchants and Company officials, especially surveyors and petty bureaucrats, were not above employing inventive and circuitous legal reasoning for their routine conquests of small patches of land, fortifying marshy spaces and establishing their *rights* and *knowledge* as the

[91] For a discussion around both the possibilities and the conceptual limits of the concept in British imperial expansion, especially with regard to Australia, see Stuart Banner, "Why Terra Nullius? Anthropology and Property Law in Early Australia," *Law and History Review* 23, no. 1 (2005). See also Michael Connor, *The Invention of Terra Nullius: Historical and Legal Fictions on the Foundation of Australia* (Sydney: Macleay Press, 2005). Legal historians and scholars have demonstrated that the concept was developed in the context of the Arctic race in the early years of the twentieth century and was anachronistically applied as a doctrine to analyze eighteenth-century European imperial expansion. Andrew Fitzmaurice, "The Genealogy of Terra Nullius," *Australian Historical Studies* 38, no. 129 (2007): 1–15. Lauren Benton and Benjamin Straumann, "Acquiring Empire by Law: From Roman Doctrine to Early Modern European Practice," *Law and History Review* 28, no. 1 (2010).

[92] See note 70.

Company gradually expanded its rule through a notarial geography.[93] We risk overlooking some of the elemental foundations of British legal architecture utilized to justify land acquisition in South Asia by ignoring questions of geography, ecology and environment. In other words, my book stresses the importance of turning to the environment for understanding law and imperial expansion by looking at how the physical properties of the landscapes, as well as the conceptual frames used to understand and measure them, impinged on local and global legal debates and the bureaucratic architecture of acquisition. Viewing urban space as an ecological terrain demonstrates how law provided the central epistemological framework in the development of imperial concepts of urban land and landed property throughout the nineteenth century.

ENVIRONMENT AND PROPERTY

This book reads the history of draining this delta as an intertwined history in the creation of a property market and a particular legal spatiality. The landscape that this book recovers can be viewed as one of the most engineered and designed "natural" landscapes in the world. Anthropogenic canals dotted this landscape prior to the coming of the Europeans, after which the colonial and later postcolonial nation-states undertook projects of reconstituting the river–land relation primarily through a propertied architecture.

While scholars have explored the making of what is now understood as "envirotechnical spaces" that involved significant engineering of river systems through dams and water control projects, this book expands that concept to look at how the property market became embedded in the technologies that built and designed this stretch of the Bengal Delta.[94] What were the environmental effects of propertizing land and waterscapes? Just as the slow movements of land and water shaped property law, similarly the expanding market in land from the middle decades of the nineteenth century onward dried, reshaped and to an extent forgot the

[93] Daniel Lord Smail proposes the concept of notarial cartography in his discussion on late medieval French spatial imagination in *Imaginary Cartography: Possession and Identity in Late-Medieval Marseille* (Ithaca, NY: Cornell University Press, 2000), 69–94.

[94] Rohan D'Souza, *Drowned and Dammed: Colonial Capitalism, and Flood Control in Eastern India* (New Delhi: Oxford University Press, 2006); Richard White, *The Organic Machine* (New York, NY: Hill and Wang, 1995); Sara B. Pritchard, *Confluence: The Nature of Technology and the Remaking of the Rhône* (Cambridge, MA: Harvard University Press, 2011).

delta. Unlike dams and multipurpose river projects, the creation of a propertied geography did not take place in a grand, technocratic manner, but was instead a slow process of accretions, erosions and sedimentations bolstered by a set of legal denominations and technological fortifications such as roads, harbors, riverbanks and a drained landscape. At the same time, this book does not merely read built spaces as sites of contested negotiations between the cultural idioms of the colonizer and insurgent rhetoric of the colonized. Instead, by introducing the work of property and its law as technology in the production of this landscape as an envirotechnical space, it becomes possible to read the slow and imperceptible reconstitution of the very specific geography of char, sandbars, marshes and bogs in the delta into an urbanized land market. More importantly, it also reveals a landscape that has sometimes resisted and often overwhelmed technology – sometimes dramatically, and at other times with gradual determination.

The particular historical ecology of this landscape is covered with layers of infrastructure "thick with power," politics, bureaucracy, speculation, legal maneuvering and cultural practices.[95] While the agrarian history of fixing the soaking ecologies is replete with multiple modalities and contestations of revenue calibration dictated by the moving landscape, the same procedures could not be transplanted to the urban spaces.[96] When agrarian land disappeared with no new land formations for five consecutive years, only then would the settlement officials recalculate the revenue drawn from that estate to reflect the disappeared land in the revenue rolls. However, the consequences of the disappearance of built land played out very differently as we see in Chapter 1 when the mobility of the landscape renders Benjamin Lacam's harbor unusable. We know regrettably little about the built landscape and the transformation of marshes into urban property and the process that this transformation set into motion.

Many of the chapters in the book look at moments of breakdown in the colonial legal logic, economic rationale and technological projects. These moments of breakdown of infrastructure and the conceptual parameters of law signal productive sites for the historian to understand the various arguments about nature, environment and technology as

[95] I borrow "thick with power" from Wiebe E. Bijker, "Dikes and Dams, Thick with Politics," *Isis* 98, no. 1 (2007): 109–23.

[96] On the calibration of revenue calculation based on changes in the course of the river, see Iqbal, *The Bengal Delta*, 18–38; Sinha, "Fluvial Landscape and the State."

articulated by the people involved. Wilson's study of Bengal's agrarian world points to the semantic difficulties in translating the property relations, which Company officials encountered among their "unfamiliar subject population," into the abstract language of liberalism.[97] While Wilson's analysis focuses on the semantic and conceptual breakdowns, my archival work pointed me to the material and infrastructural breakdowns, which became sites for reinterpreting the "natural" world of soil, land and water.

My attempt at focusing on these specific moments of spectacular and mundane failure of the East India Company, and later the British Raj, is not to argue for the limits of colonial power or for a mechanistic triumph of nature over man. These moments of failure reveal much more about the operations of colonial power and its modalities of readjusting the wheels in the face of obstruction than about *why* the projects failed. Many of these failures emerge in the colonial archive as disputes around property, forms of ownership and stakes in the land market. By analyzing how they interpreted the washing away of a harbor or the impossibility of measuring the mobile landscape, I am able to recover the contestations of economic and legal ideologies and meaning-making practices in the fluvial zone. This fluid history documents the everyday and managerial forms of British colonial conquest on this deltaic space, a process involving land, water and humans, as well as the stories and struggles that linked these various entities together.

Moreover, some of the landscapes I discovered in the colonial archive challenged all notions of hard lines separating land from water, city from river, rural from urban, private from common property. I did not encounter the well-known narratives of colonial fixities within the archive. Instead, what I stumbled on were flows, movements and circulations. The historical narrative that I unveil is a network of spaces related through law, market and engineering. Sometimes the legal and engineered geographies are layered and knotted into one another.

Too often, following the annual devastating floods in the Gangetic plain stretching from eastern India to Bangladesh, we blame the dams and embankments hastily conceived and greedily executed by the late colonial and the postcolonial nation-states. Yet the propertizing impulses of the Company that had begun restructuring these landscapes from the eighteenth century on should precede that conversation. The unique

[97] Wilson, *Domination of Strangers*, 9–13 and 45–74.

powers of the Company and the ever-shifting nature of its sovereign practices bring to light a particular legal geography that manufactured the delta as an environmental, infrastructural and built landscape.

My book is about the life of the delta at the intersection of a powerful riverine environment and imperial human intentions. While the chapters of this book follow one another chronologically, they are primarily thematically organized, and tend to go back and forth and overlap temporally. Accordingly, the organizing principle in the chapters is to analyze moments of failed legal and technological interventions within these marshy landscapes, and explain the embedded histories in which these breakdowns must be read. The book is divided into three sections. Chapters 1 and 2 comprise the first section, which focuses on the environmental consolidation of this space from the late eighteenth to the latter half of the nineteenth century. The second section, consisting of Chapters 3 and 4, explores the legal maneuvers that followed from the early attempts at fixing this landscape during the same period. The final section, containing Chapter 5 and the Conclusion, links this history of environmental and legal consolidation to the emerging market in urban land beginning from the turn of the twentieth century.

In Chapter 1 I explore how the depth of the water in the Bengal Delta was in seasonal flux, which thwarted attempts at cadastral surveys, produced inaccurate navigational charts and caused a speculative harbor construction scheme to be born. Benjamin Lacam's failed infrastructural venture to connect Calcutta to the delta brings to light the entangled relation between property, environment and law. Lacam's project ended as a court case lasting a quarter-century, a case in which the legal debate proceeded primarily along three lines: (1) the question of whether geography can serve as effective legal evidence, (2) ecological variability and the limits of hydrographical instruments and (3) human integrity in an inscrutable and mobile landscape. The controversy surrounding Lacam's harbor precipitated river and land surveys from 1790 to 1820 and resulted in the formation of new legal categories to arbitrate in this ecologically unstable landscape. This new legal framing of land and river meant that these spaces, perceived as indeterminate, came to be governed by the "emergency" provisions embedded within colonial law.

Chapter 2 turns to how the Territories and Revenue Department of Calcutta responded to the shifting landscapes. The Territories Department initiated riverbank and property surveys and devised ways to manage the tidal riparian environment by creating two new legal categories of land: new depositions became "accreted land," and disappeared land entered law as "derelicted land." This chapter contends that, from the 1820s on, the politics of acquiring and zoning land for particular purposes was embedded in the process of producing knowledge about the discrete natural and geographical elements of bogs, marshes, reeds and grass that made up the topography of this area. By analyzing the reports produced by the Territories Department of the East India Company from 1820 to 1860, this chapter argues that the newly developed legal categories engineered this landscape by separating land from water, drying the oozy mud and harnessing the marshes to create a well-defined riverbank along Calcutta. In foregrounding the surveyors, engineers, petitioners, municipal bureaucrats and proprietors, this chapter departs from earlier scholarship that has explored colonial infrastructural projects as technologies of disciplining the population. Instead, it depicts the story of the colonial struggle to transform the accretions into a stable riverbank by developing and sustaining legal fixity with paper documentation, surveys and maps that the officials hoped would reorder the architecture of ownership.

Producing Company lands in the newly formed riverbank fomented a slew of cases as many residents disputed the rulings of the Territories Department, including – perhaps most surprisingly – a Hindu deity who served as plaintiff before the British courts. The focus of Chapter 3 is on analyzing multiple cases filed in reaction to municipal projects of riverbank fortification. The shifting course of the river and the deposition of alluvion along Calcutta's western banks unfolded as two forms of crisis for the colonial government: infrastructural dilemmas, as the earlier chapter argued, and debates around ownership of the new land. Accordingly, this chapter documents the slippery terrain of ever-changing justificatory tools to ground proclamations of quasi-eminent domain in colonial jurisprudence. This laid the groundwork for a particular form of territorialized sovereignty, opening fresh questions within the scholarly debates around laws of colonial occupation.

By the 1860s, extensive parts of the delta had been drained and legally zoned to create dry docks, wet docks, urban parks and riverfront thoroughfares. Chapter 4 ties these infrastructural projects to the fiscal policies of empire, which changed in response to the evolving political situation in colonial India and global economic processes from the middle of the nineteenth century onward. In 1860, the India Income Tax Act fiscally restructured the urban economic geography of the British Raj with the introduction of income taxes, port taxes, property taxes and charitable property taxes. Reading the projects of draining as intimately tied to the fiscal reordering of the urban geography, Chapter 4 explores how new economies of value began to consolidate and restructure the already changing urban land market. These new instruments of taxation and regularization of urban landholding constituted experiments in managing the ecological spaces through law and new forms of market governance that facilitated a particular kind of growth in urban developmental projects beginning in the late nineteenth century. It is from this moment that we see a city emerge completely cut off from its hydrological ecology.

The final chapter documents the forgetting of the city's watery origin to argue that, in the dried and fixed landscape of the twentieth century, new actors emerged, muddying the lines between older and newer authorities and modes of governance. Taxation and intensification of infrastructural projects resulted in a property speculation crisis in Calcutta and across the cities of colonial India. Focusing on twentieth-century rent speculation, I track the emergence of three new political actors who began to dominate the twentieth-century urban landscape: "hoarders," planners and land developers. Novel forms of market management emerged in response to the new characteristics within the urban geography. Analyzing how these actors negotiated the land market, I demonstrate how private capital restructured the urban landscape and its marshy hinterlands. This robust and speculative market in urban land ultimately created novel networks of power with altered developmental relations to spaces within the city and its surrounding waterscape.

The management of the delta was also a management of the way we relate to marshes and remember them. The construction boom, which began in the early decades of the twentieth century, is intricately linked with the rapidly vanishing coastline of the Bengal Delta. To this day, the annual cycle of floods that drown the low-lying parts of Bengal and Bangladesh manifest themselves as disasters when the river "violates" the human-made demarcations between land and water.[98] The Conclusion brings us back to the historical present, where climate-change scientists, policymakers and the local governments of Bengal and Bangladesh are grappling with these disasters – even while rapacious real-estate developers fight for their own form of "atmosphere" by producing a dry culture in land where no land existed. The creation and consolidation

[98] Turning to the idea of a "rain-terrain," rather than the bounded world of river, Mathur and da Cunha argue, helps us recover alternative practices of design, and I would add law, too. *Design in the Terrain of Water* (Philadelphia, PA: University of Pennsylvania Press, 2014).

of a real-estate market transformed the ecology as well as our knowledge systems about the delta. The consequence of the processes initiated in the eighteenth century is becoming apparent today as the ecologically critical wetlands and floodplains surrounding contemporary Kolkata are turned into prime real estate. Newly emergent environmental movements fail to remain alert to the almanac forms of viewing spaces. Instead of seeing a mobile, elastic delta, they remain focused on artificial urban parks created by colonial drainage ventures in the nineteenth century. In a dried delta, urban parks signal value as speculative real-estate futures supplant the ecologically critical wetlands. This book tells the story of how these momentary sedimentations within the tides of the river troubled, but simultaneously solidified, multiple contradictions around ownership and land speculation over a period of almost 150 years. It reminds us of the fluid beginnings of our cities that we forgot.

PART I

ENVIRONMENTAL CONSOLIDATIONS

1

Power and Silt

> CALCUTTA
> Me the sea-captain loved, the river built
> Wealth sought, and kings adventured life to hold
> Hail England, I am Asia, power on silt,
> Death in my hands, but gold!
>
> Rudyard Kipling, *The Song of Cities*

Gold – this is what the eighteenth-century monsoon traders often called the silty strip of land they saw as their ships approached the littoral coasts of the Bay of Bengal. One may debate whether it was a metonymy for the wealth hidden in the rich alluvium or simply the result of a common mirage created by the mangrove ecosystem. This silt, responsible for creating the lower Bengal Delta, will be the starting point of this book. There are many routes to understanding the genealogy of property in colonial India. One obvious point of entry is through the intellectual history of property regimes, while another one would be the treatment of property claims in court cases. Avoiding these common approaches, this chapter will take silt – or, to be more precise, the intractable soil–water admixture peculiar to the Bengal tidal basin – for understanding the material and social processes of making property in a swamp. Silt, as Kipiling notes, is imbued with power and destruction, terms that characterize the fate of one of the East India Company's merchants in eighteenth-century Calcutta.

Nearly one-third of colonial Calcutta was still a swamp during this period. The swamps that became landed property through the nineteenth-century straddled a shifting terrain, both metaphorically and materially. The materiality of these spaces challenged the "solid" grounds for infrastructure and design, while metaphorically these swamps revealed the limits of Company's laws. This chapter explores a moment when one of the early colonial attempts to construct a harbor for Calcutta in the eighteenth century failed. This failed harbor is important for understanding how the British managed the ecology of the subcontinent over the next century and a half, restructuring the littoral landscape, shaping laws about urban property and creating a market in urban land. Following the failed attempt, the British officials turned to legal responses instead of technological adaptation to manage these spaces. This raises the following question: How did the mobility of the silty coast govern and organize the conversations around law and design in the colonial and metropolitan courts? In analyzing the terms through which the Mayor's Court in Calcutta and the House of Commons in London debated the case of the failed harbor, this chapter pursues the following line of inquiry: Why and how did law become the basis for managing these spaces? What were the conditions that enabled the legal straitjacketing of what was considered geographical difference in the "tropics" and ecological variability of tidal deltas?[1]

During the latter half of the eighteenth century, an erstwhile British draftsman-turned-private-merchant named Benjamin Lacam was discovering all the possibilities that were embedded in the malleability of the silty coasts at the edge of the Bengal Delta. He was sure that the silt the tidal Hooghly River, brought down to the flatlands forming coasts along the delta on its way to the Bay of Bengal, held a financial windfall for him. He was not wrong in his judgment, though his economically and ecologically speculative project to operationalize his harbor never saw the light of day. As the years went by, the vagaries of the silt and tides that make the delta turned his fortunes around, embroiling him in a thirty-year legal battle from which he barely escaped the debtor's prison and also lost his honor and investments in Bengal.

LEGISLATING THE VAGARIES OF SILT

Benjamin Lacam arrived in Calcutta in 1760. Thirteen years after establishing himself in various roles as a draftsman in the new construction

[1] I use the term "tropics" to signal what David Arnold calls the colonial anxiety of understanding the Indian landscape through the eighteenth century. David Arnold, *The Tropics and the Traveling Gaze: India, Landscape, and Science, 1800–1856* (Seattle, WA: University of Washington Press, 2006).

work in Calcutta, he made his way down to the mouth of the delta to the newly acquired farms of the South 24 Parganas, a territory that included the mangrove forests of the Sundarbans. The silt brought down by the Ganga-Brahmaputra river system on the eastern part of the subcontinent was already known as one of the most fertile soils in the entire world. While the mangrove areas were known to be harder to penetrate, there are records of early land grants in the Sundarban area dating as far back as the eleventh century. During the sixteenth century under the Mughals, Sufi saints from various lineages carried out extensive land reclamation and cleared forests for wet-rice cultivation.[2] On acquiring these lands, the British saw this area as uncharted and unmapped forests and wastelands waiting to be reclaimed and brought under the Company's revenue rolls.[3] In 1774, Lacam, who had not yet acquired the ranks of the coveted tax collectors of the East India Company's newly stolen farm, ventured into the tidal waters on the east of the Hooghly River to Baratulla Creek to seek his fortune in a different manner.[4] Here, he found a small patch of land that he claimed to be a "naturally forming harbour" and sought a grant from the East India Company with the purpose of building wharves and dry docks.

In this silt-laden riverine place Lacam speculated on a scheme for capital, wealth and power – all of which he had scant access to during his brief stint in Calcutta. Why did the proposition of building a harbor seem like a lucrative venture for Lacam, when the well-known practice was to try one's hand as a tax farmer?[5] In Figure 1.1, a rare trilingual hand-drawn map (c. 1730) from Frederick Barlow's collections (more on him later), we see the Hooghly River whose western banks were fast

2 For an account of eleventh-century archeological evidence, see Sayanti Pal, "Religions Patronage in the Land Grant Charters of Early Bengal (Fifth–Thirteenth Century)," *Indian Historical Review* 41, no. 2 (2014): 185–205. Pal cites two copper plates for religious land grants. For a history of the settlement of the Sundarban area, see also Richard Eaton, *Rise of Islam and the Bengal Frontier*, 1204–1760 (Berkeley, CA: University of California Press, 1993), 207–11. For a general history of the settlement of Sundarban, see Manindranath Jana, *Sundarbaner Samaj o Sanskriti* (Kalikata: Dipali Book House, 1984).

3 Sutapa Chatterjee Sarkar, "Bengal's Southern Frontier, 1757 to 1948," *Studies in History* 28, no. 1(2012): 69–97.

4 Baratulla is an eastern outlet of the Hooghly River passing east of the Sagar Islands (Saugor Island). It is also known as Channel Creek in British documents.

5 Some of the major narratives of eighteenth-century fortune-makers of the Company are about merchants, ship builders and tax collectors. Peter J. Marshall, *East Indian Fortunes: The British in Bengal in the Eighteenth Century* (Oxford: Clarendon Press, 1976).

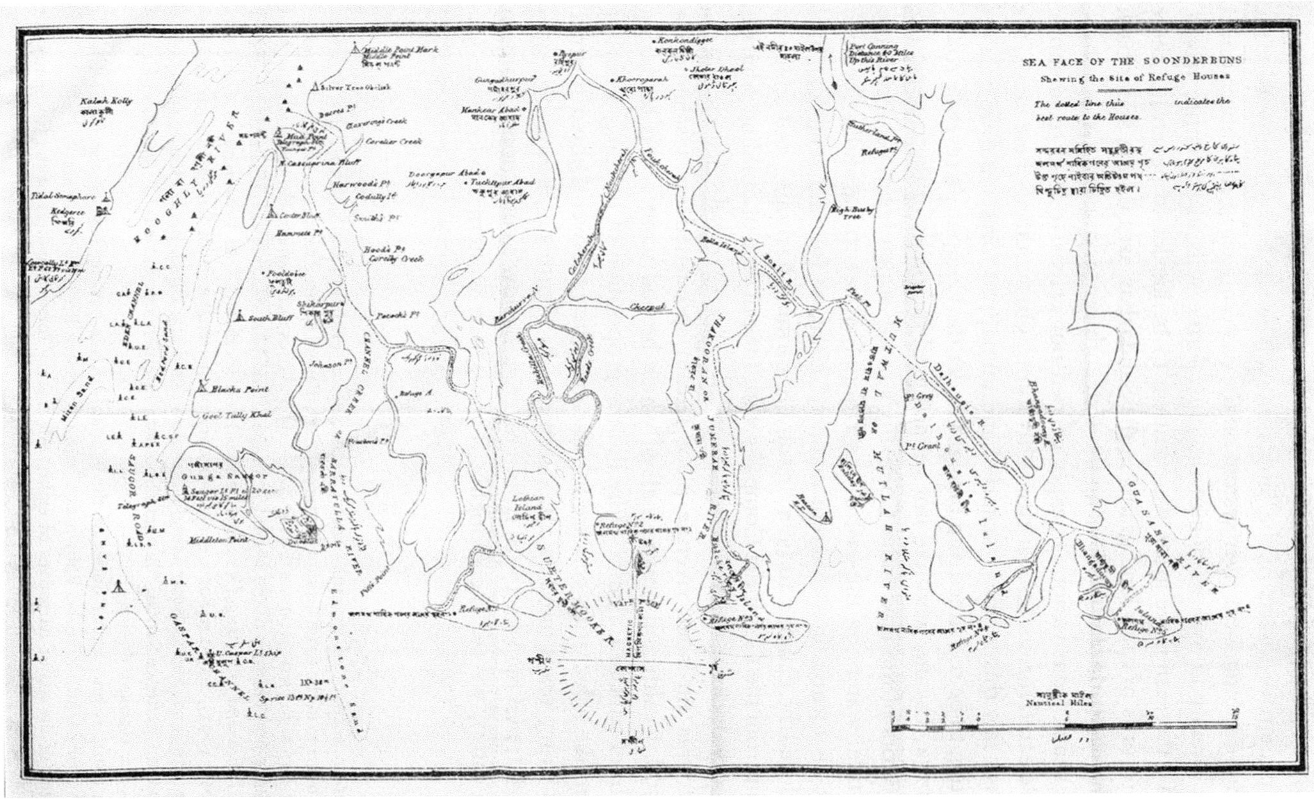

FIGURE 1.1 Map showing the sea face of the Sundarbans
© British Library Board IOR/X/9128.

silting up during the eighteenth century.[6] On the eastern side was the narrower Channel Creek, flanking a small sand-head known as Sagar (Saugor) Island. This is where the European boats left the bay to navigate their way to Calcutta through the inland riverine channels.[7] Larger ships would often need pilot boats, better adapted to tides, shoals and shifts in the river's movements, to navigate the European merchants and their goods back and forth from Calcutta.

The shipping and country boat industries were booming in the Bengal Delta through the eighteenth century. Kedgeree (or Khejuri, in the top left corner in Figure 1.1), one of the major anchorages in the Hooghly side of the river handling the highest amount of trade, was fast silting up.[8] Every mariner and ship jobber was familiar with the vagaries of the Hooghly River and the treacherous entry into Calcutta.[9] The merchants of the East India Company were looking for an alternative, and Lacam, who had already spent thirteen years in service in Calcutta, was aware of the need for a different entry. The turbulent waters of the tidal river, cyclones and the raging Bay of Bengal had sunk many a ship and a great deal of

[6] F. Barlow Collection, IOR/X/ 9128 (British Library, London). Many of the maps in the Barlow collections are undated and some are hand-sketched by him from memory. Judging from the quality of the paper, print and the cartographic style, and corroborating it with the dated maps in his collection, and other contemporaneous cartographic collections one may surmise that this map is from 1730 or afterward. Similar maps in the collection that Barlow reproduced are from 1734, although he claims that some of the them were first drafted around 1690. If one corroborates the positions with the nautical accounts in the Dalrymple Collection, this map appears to be from 1730. See Alexander Dalrymple, *Collection of Nautical Papers Concerning the Bay of Bengal Published at the Charge of the East India Company from the Mss. by Dalrymple 1784* (London: G. Biggs: British Library, 1785)

[7] There are countless accounts of Europeans sighting land at the Sagar Islands before entering Calcutta, or being piloted through the river from the Sagar Islands. See Maria Graham, *Journal of a Residence in India* (Edinburgh: Archibald Constable and Company, 1812), 129–32. Captain Sherwood and F. J. Harvey Darton, *The Life and Times of Mrs. Sherwood (1775–1851): From the Diaries of Captain and Mrs. Sherwood* (London: Wells Gardner, Darton, 1910), 360–61.

[8] Mahendranath Karan, "Khejuri Bandar," *Masik Basumati* 2, no. 4 (1925): 275–482.

[9] Surveys by French and Dutch pilots from 1748 document the existence of sandheads and silting up of the river. However, these river surveys were not accompanied by tide measurements. By 1760 Khejuri was so silted up that some ships had difficulty making the passage. An alternative passage through Channel Creek was proposed based on a survey undertaken by Major Plaisted. See James Long, *Selections from Unpublished Records of Government for the Years 1748 to 1767 Inclusive, Relating Mainly to the Social Condition of Bengal; with a Map of Calcutta in 1784*, vol. 1 (Calcutta: Superintendent of Government Printing, 1869), xxxii–xxxiii.

precious trading material before reaching Calcutta.[10] Lacam proposed to build a harbor in the eastern side, on Channel Creek. His harbor was about twenty-six miles inland from the Bay of Bengal, and it would be the closest to Calcutta.

In 1774, the Revenue Committee of the East India Company in Calcutta granted him a patch of land to begin work. Lacam invested his money in constructing what he christened the New Harbour and hoped to profit from levying tolls from the boats and ships that would be docked there. Such private investments in public infrastructure were common not only during the Mughal period in India, but also in contemporary England.[11] Yet, two months prior to completion in 1776, the Company decided to withdraw the land grant for a number of reasons, the primary of them being the contradictory reporting of water depth. Financially ruined, Lacam returned to Calcutta to file a lawsuit in the Mayor's Court. An almost thirty-year-long case began in an attempt to parse out the vagaries of tidal movements from the foibles of human avarice. No traces of Lacam's New Harbour exist today, although it was supposed to be the solution to the flux that defined the mouth of the delta.

This chapter deals with the legal case filed by Benjamin Lacam in 1777, first with the Mayor's Court in Calcutta, later appealed to the House of Commons in London and finally settled in 1806. In his complaint Lacam alleged that the Company had cheated him of anticipated profits on his investment by canceling his grant illegitimately. The Company, on the other hand, refuted his charges by arguing that Lacam's geographical measurements did not match geological facts on the ground, therefore

[10] A major cyclone in 1737 sunk eight of nine ships in the harbor. See Harry Evan Auguste Cotton, *Calcutta Old and New* (Calcutta: Thacker and Spinck, 1907), 83. The company calculated that between 1760 and 1796 it lost 20 percent of its ships to shipwreck on their way to Asia. While this may appear to be a staggering number of losses, given the existing maritime conditions and the number of sunk ships of the Company preceding 1760s, 20 percent signaled improved conditions. This number is not specific to the Bay of Bengal, but also includes wrecks in the other commonly understood treacherous parts of the route, including the English Channel, Cape of Good Hope, False Bay at the mouth of South Africa, and of course the entry to Hooghly. Shoals were often considered one of the biggest threats. It was in response to these perilous journeys as the ship entered the mouths of estuaries or bays that Alexander Darlymple was employed as the first hydrographer of the Company to create a set of printed plans and charts for the commanders of East Indiamen. See Huw V. Bowen, John J. McAleer and Robert J. Blyth, eds., *Monsoon Traders: The Maritime World of the East India Company* (London: Scala, 2011), 118.

[11] The Wells Harbor in Norfolk, which also ended up in the courts in 1782, is a case in point. See Tal Golan, *Laws of Men and Laws of Nature* (Cambridge, MA: Harvard University Press, 2009), 5–51.

raising questions about the economic viability of his plans. How is this story about the rise and fall of a middling character like Lacam, a company servant and speculator, related to the history of the emerging market in land, construction and the speculative character of that market? What does Lacam's failed harbor-building venture tell us about the changing nature of European geographical knowledge, geological debates and how those were written into the legal architecture around possession that was being crafted in Britain's eastern colony?

If the colonial riverine geography of the Iberian Empire became a site to imagine political dangers, temptations for self-rule made possible by the coastal corridors and ultimately imperial territorialization, then a different history emerges in the tidal coasts on Bengal.[12] This thirty-year legal case serves as a cipher through which we can trace the dynamic unfolding and refashioning of metropolitan law's attempt to apprehend an unknown geography and its vicissitudes in the creation of both legal structures and geographical knowledge. The legal battle surrounding Lacam's harbor reveals the beginning of a century-long process where the multiple ways of understanding and living with the ecology of the tidal basin was condensed through contractual principles of ownership and scientific instruments of knowing and measuring space. Lacam's project revolved around a new path for sailing to Calcutta from the Bay of Bengal that he claimed most sailors were not aware of. This new track had to be fortified by constructing a harbor on this mobile landscape to facilitate the movement of cargo and fleets. For the officials employed in Calcutta, the mariners who plied the region and the Company's Board of Directors in London, Lacam's legal case rested on uncovering the relation between geography, human error and law.

In arbitrating the case, the colonial officials grappled with the following questions: Was geography an unstable and an unreliable form of evidence; or was it a matter of human miscalculation, whether purported or otherwise, in order to make wealth out of mud and profit out of geographical indeterminacy? In deciphering this landscape and water bodies as it emerges in the legal debate and the remembered histories of the place, we begin to see the particular entanglements within geography, scientific developments of maritime measurements, geological sciences and law that the mobile landscape of the delta stitched together. This case also laid the groundwork for geographical formations in the tidal

[12] For this line of argument about riverine legal corridor, see Benton, *A Search for Sovereignty*, 40–103.

delta to become part of the legal, governmental and scientific authority in the empire, thus slowly transforming the relation between land, water and forms of possession that inhere in those spaces.[13]

Lacam's harbor emerges as a site for reading the consolidation of colonial power as a bureaucracy of violence. In what follows, this chapter reads the various arguments and legal reasoning used in Lacam's case, including the depositions, refutations and the legal proceedings by the company servants, merchants, sailors and builders as sets of information about what they perceived as an indecipherable ecological difference. The lawmakers in Calcutta and London reinscripted this information into a particular legal semiotics of property, with a broad-ranging impact that rearranged the space of the tidal swamp in the colony.[14] The debates about the geological features of a tidal delta and science of harbor construction became sites for producing legal testimony, arbitration around ownership disputes and compensation. This case illuminates two emerging aspects of colonial power at the turn of the eighteenth century: first, the politico-social world that the Company was establishing as an expanding imperial power in South Asia; and second, a particular colonial social network where "nature," "ecology" and "property" all emerge as historically contingent categories created in and through the debates and discussions between the lawmakers, company servants, petitioners and mariners. As we will see, each of these groups had different ways of reading the material landscape.[15]

GEOGRAPHY AS EVIDENCE

In 1760, at the height of the seven-year Anglo-French War (1756–63), or what was known as the Carnatic War in India, Lacam set forth on a ship from Madras to Bengal with the ambition of making a fortune. He was

[13] Michel Foucault, *The Order of Things: An Archaeology of the Human Sciences* (New York, NY: Vintage Books, 1973), 212–18.

[14] Anthropological scholarship as well as science and technology studies have read law as material practice and embedded in a network of association that it creates in the process of forging those associations. Bruno Latour, *The Making of Law: An Ethnography of the Conseil D'etat* (Cambridge: Polity, 2010); Annelise Riles, *Documents: Artifacts of Modern Knowledge* (Ann Arbor, MI: University of Michigan Press, 2006); David Cowan and Daniel Wincott eds., *Exploring the "Legal" in Socio-Legal Studies* (London: Palgrave Macmillan, 2016); Cornelia Vismann and Geoffrey Winthrop-Young, *Files: Law and Media Technology* (Stanford, CA: Stanford University Press, 2008).

[15] Alain Pottage, "The Materiality of What?," *Journal of Law and Society* 39, no. 1 (2012): 167–83, 171.

skilled in drawing and was armed with a recommendation to Henry Vansittart, the then Governor-General of Bengal. On his arrival in Calcutta, Vansittart first appointed Lacam to be an assistant to his secretary's office, where Lacam's drawing skills brought him recognition. From there he was promoted to the position of Assistant to the Chief Engineer of Fortification and New Works in Bengal. Lacam's ambitions proved too wide for his narrow position as an assistant and draftsman, and he took charge of the musters, volunteered to clerk for the treasurer's office and set about diligently detecting fraud and discrepancy in the new construction works in Bengal. Construction, it seems, was always beset with corruption. Lacam left a mark with the Court of Directors of the East India Company because of his assiduousness in detecting and eliminating corrupt practices in the office.[16]

Lacam, who now considered himself on the path to amassing greater fortune than his office as the Muster Master of the treasury allowed him, left Calcutta soon thereafter. He set forth to the Sundarban area, a mangrove forest at the entrance to the Bay of Bengal and the mouth of the Hooghly River. Marshy tracts, dense undergrowth, saline creeks and a whole host of wildlife, from tigers, to crocodiles to a rich bounty of fish, populate this region. Life and livelihood in this area is shaped by a very specific relation that binds humans with animals in the soaking ecology of the Sundarban. Tiger gods and river spirits appear as mythic figurations and constitutive elements structuring the politico-ethical world in the land of mangrove and tides.[17] Even in the latter half of the eighteenth century, much of the area was neither charted nor mapped by the British, although the Mughals settled substantial tracts in the Sundarban area during the sixteenth century.[18]

[16] Papers concerning Benjamin Lacam and the New Harbour in Bengal, IOR/H/Misc/396: 1765–1809, 7–10 (hereafter Lacam's Private Papers).

[17] It is beyond the scope of this project to delve into this, but for an understanding on how life among animals has shaped livelihood in this area, see Pradip Kumar Barman, *Baghera Rajatve: Sundarabana Bishayaka Galpasamkalana* (Mayahari: Samakaler Jiyanakathi Prakashana, 2012); S. K. Pramanik, *Sundarabana: Jala, Jangala, Jibana* (Calcutta: Paribeshaka Pustaka Bipaṇi, 2008); For two anthropological accounts and the importance of the tiger myths, see Annu Jalais, *Forest of Tigers: People, Politics and Environment in the Sundarbans* (New Delhi: Routledge, 2010); Chandan Surabhi Das and Sunando Bandyopadhyay, *Sharing Space: Human-Animal Conflicts in Indian Sundarban* (Kolkata: Progressive Publishers, 2012).

[18] Abul Fazal's *Ain-i-Akbari* mentions villages such as Hathiagarh, Mednimal (near modern-day Canning), Maihati and Dhuliapur, which currently dot the northernmost borders of the area.

Although the Company acquired the rights to farm this area in 1757, it soon discovered that property in land and silty riverine spaces was less about contracts and productive control and much more about obligations, gifts, social capital and political power.[19] Initially, they were unable to make any serious efforts to survey the area.[20] With much difficulty, the Company was able to secure the services of Captain Robert Barker and Captain William Swallow and a huge retinue of *kanungo* (registrars of a fiscal division), who set out to "prepare a plan of the course of the Great Lakes, sound its depths of water, etc."[21] Sounding was a particular method to measure depth of water as well as detect the presence of shoals, sandheads and barriers in rivers and oceans by using a heavy piece of lead attached to a line. According to their survey, the *zamindars* (landlords) had over 816,446 *bighas* of land, and they collected rent on 454,804 bighas, just half of the land in their possession.[22] The rest of the land was "either barren and untenanted, or assigned over to servants, idols, etc."[23] The Company soon realized that land assessment in this area required specialized knowledge and decided to farm out the lands to people who possessed those skills. A concerted effort at reclamation and mapping began much later under the Collector General Claude Russell in 1770.[24]

Thus, for the initial sixteen months following the acquisition of the South 24 Parganas in 1757, the Company collected revenues and

[19] Historians of southern India have explored this in length, especially with regard to temple properties. See Pamela G. Price, "Raja-Dharma in Ramnad, Land Litigation, Largess," *Contributions to Indian Sociology* 13, no. 2 (July–December 1979), and also her *Kingship and Political Practice in Colonial India* (Cambridge: Cambridge University Press, 1996); Nicholas B. Dirks, "From Little King to Landlord: Property, Law and the Gift under the Madras Permanent Settlement," *Comparative Studies in Society and History* 28, no. 2 (1986): 307–33, 324. For a study on property and Sufi shrines, see Richard Eaton, "The Political and Religious Authority of the Shrine of Baba Farid in Pakpattan, Punjab," in *Moral Conduct and Authority: The Place of Adab in South Asian Islam*, ed. Barbara Metcalf (Berkeley, CA: University of California Press, 1984), 333–56.

[20] Frederick E. Pargiter and Frank D. Ascoli, *A Revenue History of the Sundarbans*, 2 vols. (Kolkata: West Bengal District Gazetters, Govt. of West Bengal, Higher Education Dept., 2002).

[21] Entry 887, August 4, 1757, in *The Press List of the Ancient Documents Preserved in the Imperial Record Room of the Government of India* 1757–1758, vol. V (Calcutta: Superintendent Government Printing, 1899), 63–66.

[22] Bigha is a unit of land measurement prevalent in parts of South Asia. In Bengal bigha was standardized by the British as 1 bigha being equivalent to roughly one-third of an acre.

[23] Walter Kelly Ferminger, "The English Acquire the Twenty-Four Parganahs," in *The Fifth Report from the Select Committee of the House of Commons on the East India Company, Dated 28th July, 1812*, vol. 1 (Calcutta: R. Cambray & Co. 1917), xcix.

[24] Chatterjee Sarkar, "Bengal's Southern Frontier," 72.

thereafter leased the lands to the highest bidder on three-year terms. The Indian zamindars claimed those lands, water, mangroves and forests as their own, and some came forward with high bids to the Company. Although the Company wanted to farm the lands itself, Zephania Holwell, the tax collector and self-styled zamindar of Calcutta, worried that keeping the lands such as those on the silty riverine delta in the inexperienced hands of the Company would inhibit them from discovering the real value of land.[25] Giving it to the Indians, he surmised, would produce a "private confederacy" in those lands that would be "injurious" to the Company's interest.[26] Unlike the Jessore or Bakarganj districts, much of which was also mangrove forest and settled as *patitabadi taluks* (waste or fallow land), in the case of the 24 Parganas, the Company wanted to take a different approach and keep the area under their direct control, especially given its proximity to the various ports and Calcutta. Ultimately, that was not possible.

The "conundrum" of assessing the extent and value of the area, however, was not unique to just the mangrove forests. Even the Permanent Settlement Act of 1793 did not properly solve the Company's problem with land assessment. Rather than a clear demarcation of proprietary rights, landed property traversed a flexible terrain of political deals and negotiations.[27] In the meantime, the Company divided the entire area into fifteen parcels. Through a public auction they leased these parcels to indigenous merchants, former Indian proprietors and some company servants.[28] Benjamin Lacam was one such proprietor.

Here, in this shifting riverine entrance, Lacam's fortune and failure would be etched. He applied for a grant of land, but unlike the other proprietors, he did not wish to be an agrarian landlord.[29] Rather, he wanted to privatize a piece of the coastline and eventually hold private rights in the name of Company's public works. Unlike the other proprietors, Lacam initially received a very small tract of land of 100 bighas. Following his stint in construction in the new Fort William in Calcutta,

[25] The question of real value of land is what Travers explored in "'The Real Value of the Lands.'"

[26] Firminger, "The English Acquire the Twenty-Four Parganahs," ci.

[27] Washbrook, "Law, State and Agrarian Society in Colonial India," 664–65.

[28] For the list of proprietors, see Firminger, "The English Acquire the Twenty-Four Parganahs," ci.

[29] Surrender of Pottah, IOR/L/L/2, 1812 (British Library). The Company granted Lacam about 500 bighas of land in 1767 and he surrendered his land grant to the Company in 1791.

built by the Company following the Battle of Plassey in 1757, he decided to try his luck by initially becoming a supplier of building material. He set up a *chunam* (lime) works in the Sundarban area and became a contractor for the Company's expanding infrastructural and building works. He soon succeeded in this endeavor, supplying lime at a cheaper rate than at the existing market prices, beating his competitors who were supplying chunam from Sylhet, a city on the Surma River in northern part of Bengal.[30]

In the process of establishing himself as the lime supplier to his old office, the Department of Fortification and New Works, Lacam cleared vast tracts sixty miles downstream from the English settlement in Calcutta to make wharves for shipping lime and also for building smaller vessels (known as sloops or country boats) better adapted to inland riverine trade. His work in the Sundarban area familiarized Lacam with the vagaries of the landscape, and, entrepreneurial as he was, he soon discovered an opening for a new venture: to build a permanent harbor there. While working as a contractor of lime for the Company he began surveying the channels and making soundings of the area to fathom its depth of water at both high and low tides.[31] This was an early attempt to improve Calcutta's connection with its southward ports at the mouth of Bay.[32]

Private investment in public infrastructure might strike us as unusual today, but that was not the case in either early modern South Asia or England. During this time, *banias* (Indian bankers) who worked closely with the imperial treasury of the late Mughal Empire financed infrastructural works. Constructing ports (*bandars*) was as lucrative as owning a marketplace.[33] Ports were part of the wide network of pilgrimage, war

[30] Sylhet lime was highly regarded and used during the late eighteenth and nineteenth century for construction work in Bengal. For more on the history of lime production and its market, see T.R., "On the Manufacture of the Sylhet Lime," in *Gleanings in Science*, February 14, 1830 (Calcutta: Baptist Mission Press, 1830), 61–63.

[31] Report from the Committee to Whom the Petition of Benjamin Lacam, Esq. Grantee of New Harbour, in Bengal Was Referred, House of Commons, Great Britain, July 10, 1806. Appendix I, 16 (hereafter Petition of Lacam, HC 1806).

[32] Lacam's Private Papers, 78.

[33] Most of the area that became colonial Calcutta prior to the establishment of British settlements was farmed out as weekly markets by the landlords. For a history of these markets, see Shrimoyee Basu, "'Bazaars' in the Changing Urban Space of Early Colonial Calcutta," PhD thesis, University of Calcutta, 2015). On the importance of controlling the lease to such markets in the Indo-Gangetic plain and how that was transformed with the colonial encounter, see Sen, *Empire of Free Trade*; Anand A. Yang, *Bazaar India:*

and trade routes in the Gangetic plain, drawing revenue through tolls, taxes, ferry passes, overseas port customs (*mir bahari*), pilgrimage dues and market taxes (*zakt*). Bijayaram Sen's *Tirthamangal*, a didactic poem composed in 1777, attests to the growing importance of ports along the Gangetic plain.[34] Some of the important inland ports in seventeenth- and eighteenth-century Bengal were Shahbandar in Dacca, Bakshabandar in Hooghly and Pachotera in Murshidabad. These were the dominant ports before Khejuri, Sagar and Calcutta became the primary centers of European trade. The Nawab (King) of Bengal collected tolls through a wide array of administrative posts employed in these ports.[35] Lacam attempted to fashion his harbor along those lines with the hope of taxes from the passing vessels and rewards from the Company through an elevation of his status.[36]

With the permission from the Revenue Committee in Calcutta, Lacam began to invest his own money for the purposes of "publick works" as he defined it.[37] One of the earliest British descriptions of the site where he began work on his harbor predates Lacam's so-called discovery of the place and comes from Captain John Ritchie, who initially surveyed this area between 1767 and 1770. Ritchie described the place in the following words: "[It was] secure from every wind, and the water never being agitated any thing worth notice; for the tides in it are strong, and they are also irregular in their motions, occasioned by the sudden bending of the River at nearly right angles."[38] Borrowing from Ritchie's description, and claiming it to be his discovery, Lacam made a case to the Company that the constant ebb and flow made it a strategic harbor location. Ultimately in 1770 Lacam won a unique land grant to privatize this space in the name of public service.

How do we read Lacam's glossing over of Ritchie's earlier survey and his claim to discovery? Although a comprehensive survey of the area had yet to be conducted, Lacam's analyses were built on existing small-scale surveys. Apart from the earlier French and Dutch riverine surveys,

Markets, Society, and the Colonial State in Gangetic Bihar (Berkeley, CA: University of California Press, 1998).

34 Bijaẏarama Sen, *Tīrtha-Maṅgala* (Calcutta: Paraśapathara Prakasana, 2009).

35 Sen, *Empire of Free Trade*, 38–42.

36 Report of the Collector of Customs at Hughli on the Prevalent Mode of Operation at Native Custom-Houses Prior to the Regulation of 1773, IOR/Home Misc/216, 6–7.

37 Throughout his proposal and later in his petition, Lacam calls his New Harbour a public work undertaken with his private funds.

38 Petition of Lacam, HC 1806, Appendix 2, 24.

Captain Plaisted conducted a separate small survey in 1760. As far as land survey was concerned, Robert Barker conducted the first one with great difficulty.[39] This was followed by a statistical survey of the landholdings by William Farkland, who was appointed as the tax collector of the "New Lands," as it was called for some time.[40] There were two other surveys of this newly acquired land. Marine Surveyor Captain John Ritchie undertook a nautical survey of the land–water admixture lasting three years from 1767 to 1770.[41] The other was much more extensive in scope, conducted by Major-General James Rennell, whose survey of the lower Bengal Delta began in 1764 and was completed over a period of thirteen years. Lacam's survey, conducted on a much smaller scale and with the purpose of demonstrating the potential for Channel Creek and the surrounding area to be an ideal wharf for the Company's ships, drew tactically and selectively on Ritchie's surveys. While he borrowed and corroborated his facts with Ritchie's work, he had to exchange heated letters with Rennell to disprove Rennell's findings and establish his deductions about depth of water in this space.[42] Lacam argued his intimate experience with the landscape made him a better judge than Rennell's scientific abstractions. Such contestations between experiential and scientific knowledge were common, especially when cases such as Lacam's ended up in law courts. John Heilbron's and Tal Golan's study of the Wells Harbor case in Norfolk, which took place around the same time, documents a similar rift between the veracity of experience on the one hand and the expert claims of abstract scientific facts on the other hand.[43] Lacam claimed that his initial surveys revealed that the site had the potential to be a harbor, offering a new passage circumventing the treacherous entrance on the Hooghly to Calcutta – thereby promoting the trade

[39] While Lacam's Private Papers as well as Long's *Unpublished Records* both mention his survey, the actual survey report is untraceable.

[40] Firminger, "The English Acquire the Twenty-Four Parganahs," xcix.

[41] John Ritchie, *An Hydrographical Journal of a Cursory Survey of the Coasts and Islands in the Bay of Bengal by Capt. John Ritchie 1770 and 1771 Published from the Ms. at the Charge of the East India Company by Dalrymple* (London: G Biggs 1784).

[42] For an account accusing Lacam of lying, see also Maps of F. Barlow, IOR/X/14696, British Library, UK.

[43] I would like to thank John Heilbron for sending me his unpublished paper "The Laws of Nature and the Laws of the Land" (October 1995), which discusses the Wells Harbor Case to highlight the entanglement of natural philosophy, legal theory and theology in the early modern period. See also Golan, *Law of Men*, 30–32.

and military position of the town. He further added that his personal experience with the landscape was also corroborated by other surveys.[44]

Lacam was not wrong to draw on earlier surveys. Surveys during that period straddled various genres of representation and narrative, and were a collaborative effort across generations based on collection, collation and collaboration.[45] "Geographical science" was, according Thomas Best Jervis, an engineer and surveyor with the East India Company between 1819 and 1830,

> [t]he simultaneous exertions of many individuals wholly unknown to each other, to institute inquiries preparatory to that enlarged and more exact acquaintance with the relative situation of countries and objects on the surface of the globe, its precise form, dimensions, distribution, and local peculiarities; these all have followed successively at intervals, as investigations supposed to have originated in fortuitous circumstances which some one or other of the foregoing causes had contributed to elicit or suppress.[46]

This account fittingly describes the nature of the task Lacam undertook. Neither a geographer nor a trained surveyor, Lacam based his findings on earlier claims and "the knowledge he gained from long, long experience" to make his observations about the landscape where he had been toiling at his lime works for a number of years.[47] Knowledge about the swamps of Bengal was created as cases about Lacam ended up in the legal courts as a matter of jurisprudence of evidence. Facts slowly congealed around the contestations between the laws of men, including the speculators, surveyors and judges, and the laws of the inscrutable tidal swamp.

Lacam argued to the Revenue Department of the East India Company that New Harbour was a strategic location, given that it was tucked away from the unstable tides that surrounded much of the area around the factory town of Calcutta and the mouth of the Hooghly River. Moreover, Lacam stressed that the site he chose for the New Harbour was without the dangers of the shoals or sandbanks that the Hooghly

[44] Before Lacam, Plaisted had pointed out that Channel Creek could replace the previous entrance through the Khejuri part of Hooghly. At a later date, though, Lacam will try to claim this as his discovery.

[45] Paul Carter, *The Road to Botany Bay*; Lorraine Daston and Peter Galison, *Objectivity* (Cambridge, MA: MIT Press, 2007).

[46] Thomas Best Jervis, "Memoir on the Origin, Progress, and Present State of the Surveys in India," *Journal of the Royal Geographical Society of London*, vol. 7 (1837): 127–43, 129–30.

[47] "Letter to Major James Rennell from Benjamin Lacam," in Ritchie, *An Hydrographical Journal*, 59–60.

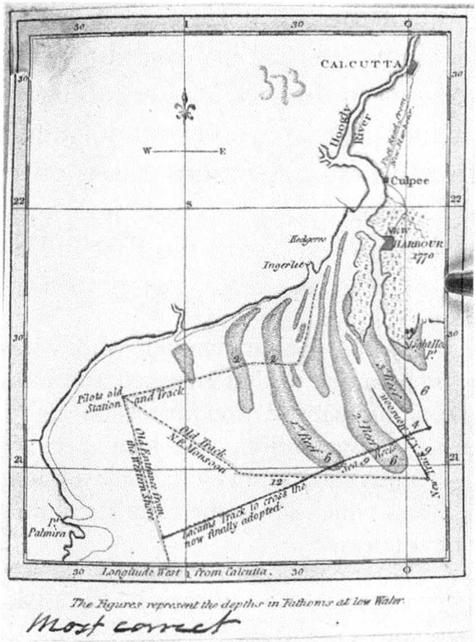

FIGURE 1.2 Map showing Benjamin Lacam's proposed routes and the New Harbour location.
© British Library Board IOR/H/ 396: 1765–1809: 321.

River disembogued annually. Commenting on the existing passage, the master-navigator Captain Ritchie previously declared that "Nature" had done little for the safety of the Hooghly River and that "[a]rt has done nothing to accommodate them, in repairing, cleaning and rebuilding." Therefore, one is left to procure large sums of money here to repair the "injuries of Time."[48] This space was a "naturally forming harbour," according to Lacam, and building the docks would afford the merchants of the company with a new route (marked as Lacam's track; see Figure 1.2), which had the double advantage of being shorter, while also avoiding the old and treacherous channel and its multiple sand banks (marked as Old Track, or Old Entrance; see Figure 1.2). He further added that the Company would gain all of this without spending any of its own money.

Many of the merchants and marine surveyors attested to the difficulties of crossing the bay into the river to get to Calcutta and to the limitations

[48] Captain Ritchie's Deposition January 1776, Lacam's Private Papers, 97–98.

of navigating the tidal basin.[49] Captain John Hicks, who had been employed in the sea service of the Company for nineteen years, during which period he commanded ten commercial and military ships, spoke in favor of Lacam's proposed venture. The difficulties of navigating the entrance of the Hooghly River arose from the fact that it was hard to calculate or expect sufficient depth to navigate large ships.[50] The capriciousness of the tidal river meant that ships could cross only during high water. Therefore, each crossing had to be accompanied by the time-consuming task of sounding the river to gauge if sufficient depth was available for the purposes of navigation. Hicks was not the only one who spoke about the dangers of the river. Captain Newte, another merchant, further added that "the water in the Old Channel [existing path] was so shoal, that the East India Company's ships, when laden, are obliged to sail for many miles in water not above a foot deeper than what the ship draws."[51] According to Captain Robert Preston, one of Lacam's colleagues, the absence of sufficient water depth meant that boats were often halted in their journeys to the ports of either Kidgree or Ingelle, little over a hundred miles from Calcutta. Not only did this cost the Company a great deal of money, but often merchant ships had to turn around and return before unloading their entire shipment in Calcutta or one of its neighboring ports.[52]

Commercial losses were not the only reason driving the necessity of a safe harbor. The presence of the Dutch and the French competition along the coast meant that the British needed harbor infrastructure and military fortification to protect their Calcutta stronghold. Captain Preston added that any passage of ship

> would be attended with the utmost difficulty and danger to attempt getting a line of battle ship into the river Hughly by the Channel now in use, which is amazingly changed, by reasons of Sands, from what it was when Admiral Watson passed it; that the risque now would be immense, for a ship above twenty feet draught to pass Ingelle.[53]

[49] Anon., *Thoughts on a Late Proposal for Piloting Ships at All Seasons through Channel Creek, Instead of the Old Passage by the Braces, into Hoogly River* (London: J. Stockdale, 1782).

[50] Petition of Lacam, HC 1806, Appendix A, 7.

[51] Petition of Lacam, HC 1806, Appendix A, 8.

[52] Petition of Lacam, HC 1806, Appendix A, 8.

[53] Petition of Lacam, HC 1806, Appendix A, 8.

This old channel was daily becoming worse, thus causing immense harm to the King's service. Given the general opinion about the treacherous nature of the existing channel and the need for an alternative route, Lacam scoped out a new venture in his harbor, an area surveyed and explored in depth, he claimed, only by him.

In the process, he advanced an infrastructural solution to the mercantile military limitations posed by the tidal river. He proposed to build his harbor with his private money and the Company's support through land grants and supplied the Company with the plan shown in Figure 1.3.[54] In the month of February 1774, he applied to the government in Fort William, Calcutta, for a land grant of 100 bighas, and within a fortnight his grant was approved at the existing rental rate for "wasteland."[55] He then began deforesting the area and displacing the coastal settlement of "wild beasts."[56] In October 1775, within a year and a half, he intimated to the government his decision to build a wharf for the Company there since this space provided a critical new track in the treacherous river where "[s]hips of any Burthen, may ride secure at all Times and Seasons, sheltered from every Wind, and in no Case subject to the Inconveniences arising from the excessive rapid Tides, known to prevail ... [in] every part of Huegley River, quite to Calcutta."[57] In the words of his supporters, the New Harbour was "a Place, entirely of his own forming."[58] It is surprising that each one of the sailors and surveyors who supported his proposal spoke of the flux in land and water in terms of anomaly: caprice, vagary, changeable and subject to the strange moods of the rivers.

There was consensus that the space was anomalous and needed to be managed. For the next year, Lacam worked on building a pier and a wharf to accommodate the largest ships of the Company. The Company supported him through further land grants and by supplying him with chains for moorings, material for his docks and gates from the stores in Calcutta to expedite the construction works. By the winter of 1776 his harbor was nearly complete. In February 1776, a month before its expected completion, he asked the Board of Directors for two things. First, he wanted them to support his application for another tract of land through a grant of four hundred bighas, substantially larger than his previous request from the Calcutta Committee of Revenue. Second, he wanted the Company's Master Attendant to survey the channel up to his

54 Lacam's Private Papers, 320.
55 Petition of Lacam, HC 1806, Appendix 1, 16.
56 Lacam's Private Papers, 78.
57 Lacam's Private Papers, 78.
58 Lacam's Private Papers, 78.

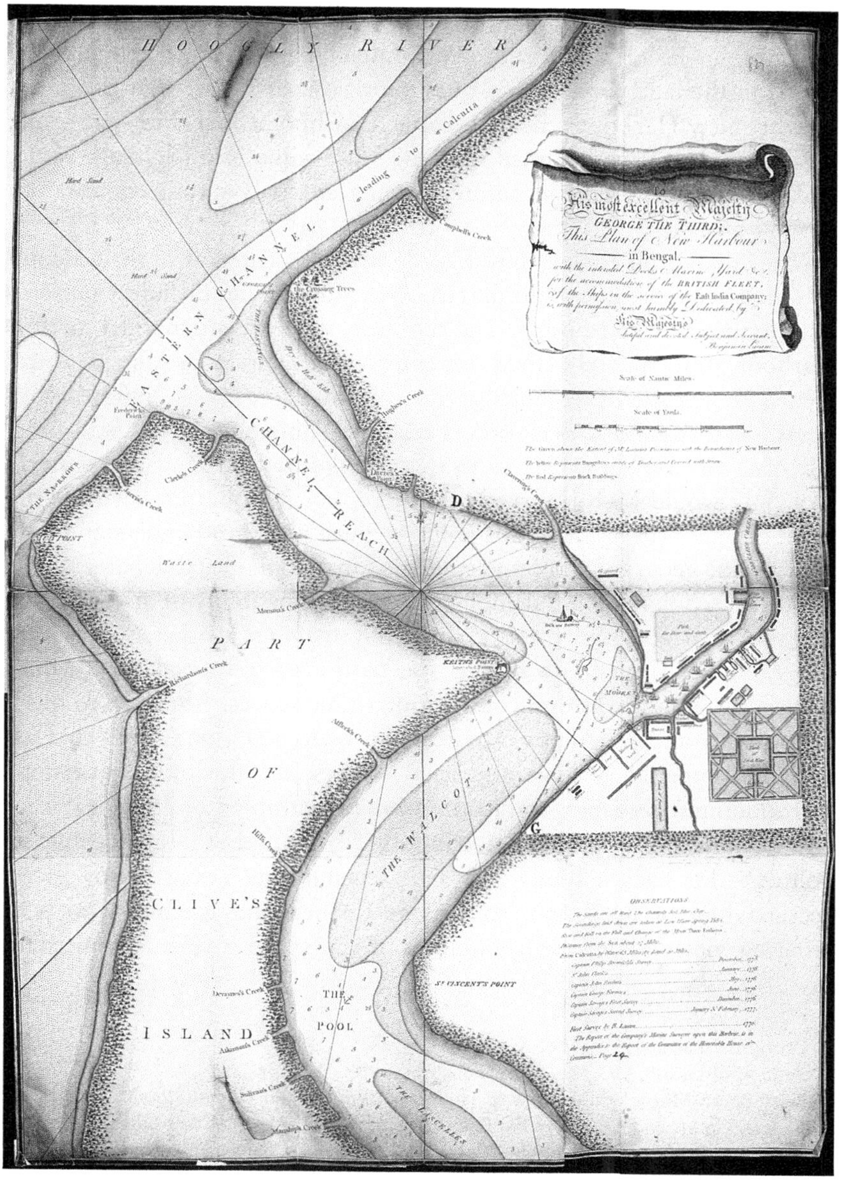

FIGURE 1.3 Map showing Benjamin Lacam's proposed harbor plan.
© British Library Board IOR/H/Misc/396: 1765–1809.

harbor "which might prevent any deception" as to ownership.[59] For the first time in his dealings with the Company, Lacam turned the rights vested in the land grant into a question of ownership. It is at this point that his New Harbour scheme became a flashpoint in a thirty-year-long legal battle, as Lacam turned the land's use-value into exchange-value, and privatized the flux in nature, turning anomaly into a contestation around profits.

The initial problem began to emerge when the Director of the Calcutta Revenue Committee realized that the *land* grant would include a substantial number of water bodies. The surveyor, who was discharged to New Harbour, unfortunately could not complete his survey or take measurements. He reported that much of the mangrove forest was "infested with tigers" and that the survey orders on this "subject respected land only; whereas Mr. Lacam's address to them spoke both of land and water." Not only would such a survey cost more, but it would go beyond "the spirit of their orders."[60] At this point we must pause and understand how this specific geography became entangled in colonial legal debates.

Recent scholarship has demonstrated the impossibility of disentangling law, imperial geography and cartography from each other.[61] The legal geography of the empire reveals the nature of territorial sovereignty, especially along the littoral coasts or maritime spaces. Attention to questions of law and geography, according to Lauren Benton, shows that for both the imperial officials and legal writers, the theoretical questions surrounding sovereignty and justice were inseparable from the local legal politics and the interimperial contests.[62] Attention to the local legal politics – for Lacam surely had many enemies, including Barlow, who accused him of deception, as well as Rennell and Ritchie, who held differing views about the geography of the space – reveals something

59 Petition of Lacam, HC 1806, Appendix 1, 17.

60 Petition of Lacam, HC 1806, Appendix 1, 17–18.

61 Gary Bryan Magee and Andrew S. Thompson, *Empire and Globalisation: Networks of People, Goods and Capital in the British World, c. 1850–1914* (Cambridge: Cambridge University Press, 2010); Benton, *A Search for Sovereignty*; Sameetah Agha and Elizabeth Kolsky, eds., *Fringes of Empire: Peoples, Places, and Spaces in Colonial India* (New Delhi: Oxford University Press, 2009); Sally Engle Merry, "McGill Convocation Address: Legal Pluralism in Practice," *McGill Law Journal* 59, no. 1 (2013): 1–8; Eliga H. Gould, "Entangled Histories, Entangled Worlds: The English-Speaking Atlantic as a Spanish Periphery," *American Historical Review* 112, no. 3 (2007): 764–86, and "Entangled Atlantic Histories: A Response from the Anglo-American Periphery," *American Historical Review* 112, no. 5 (2007): 1415–22.

62 Benton, *A Search for Sovereignty*, 5–7.

other than political void or questions of sovereignty. Both Rennell and Ritchie, surveyors of the lands and waters under the Company's jurisdiction, had political stakes in shaping the geography of the delta toward productive purposes. Geographical authority was a prized possession in imperial expansion. Apart from the disputed nature of ownership of the land, questions of geology of the tidal delta played an important role.

Benton's work has already unsettled the idea of bounded, land-based empires to reveal fractured legal geographies coexisting with unregulated spaces. This opened the space for a study of how both real and imagined spaces were products of nonmetropolitan articulations of law, legal visions and the ad hoc nature of legal reasoning and jurispractice.[63] Renisa Mawani pushes the terrestrial and maritime divide to focus on amphibian legalities, thus raising the question of how we attend to the specificities of a tidal swamp or the *char* ecology of this area. If char was understood as geographical difference, how did it affect the question of law? In the case of Bengal, this means specifically: How did physical features that remained beyond the purview of Western geographical science, including a tidal moving landscape, where land and water seemed inseparable, shape property laws and lead to a legal doctrine to settle geographical disputes?

Lacam's failed infrastructural project organized questions of property, environment and law in the colonial and metropolitan court. The geographical specificity of the Bengal Delta demands that we frame the legal question differently. This is especially true of the landscape Lacam wanted to acquire, possess and integrate into the space of Calcutta. Thus, questions of ownership in the temporary sandbanks, which were neither land nor water, had to be managed differently.[64] The concepts of delta, land, river and tides emerged as scientific concepts within the legal domain and everyday metaphors to understand geographical exceptions. These exceptions required translations within

[63] Nandini Chatterjee and Lakshmi Subramanian, "Law and the Spaces of Empire: Introduction to the Special Issue," *Journal of Colonialism and Colonial History* 15, no. 1 (2014); Iza Hussin, "Circulations of Law: Cosmopolitan Elites, Global Repertoires, Local Vernaculars," *Law and History Review* 32, no. 4 (2014): 773–95.

[64] "Moving Islands," the final chapter of Tamar Herzog's recent book, *Frontiers of Possession: Spain and Portugal in Europe and the Americas* (Cambridge, MA: Harvard University Press, 2015), explores the changing meaning-making practices within law that accompanied the natural changes and changing patterns of land use in the Iberian Empire.

law and infrastructural discussions in order to mediate metropolitan desires with the "unstable" colonial ecology.

In Lacam's case, for instance, before the soil could be understood as measurable, mappable property and a productive entity, it first had to be produced both materially and ontologically. For the soil to be abstracted into capital, it first must be recognized as productive soil. This was not always the case in eighteenth-century Bengal, a place where the arrangement of space through abstraction as capital was always interrupted by the natural forces of the fluvial landscape that challenged any pregiven ontology of the soil.[65] If we move away from the one-dimensional ways in which Lacam's plans depicted this space and remind ourselves about the almanac forms of understanding the mobile nature of this landscape, we see that soil is temporary in the lower Ganga Delta basin. What is soil today might be a floating land–water admixture tomorrow, thus challenging all existing ideas of territorial or conceptual fixity.[66] The unsteady soil enters colonial discussions by interrupting legal systems of registering land and scientific techniques of measuring water depth.

The presence of water also threw up questions of use rights. Captain Ritchie noted the question of use rights in his deposition in 1777 with regard to Lacam's land grant. While Captain Ritchie understood the relative advantages of the site, he could not address the question of whether private property could be invested in the water bodies within the granted land in the Channel Creek. He was concerned whether a land grant to Lacam "could be given without invading the common rights of mankind."[67] The problem of soil–mud–water admixture would continue to affect the workings of both the legal and the engineering departments of the Company. The Company's bureaucratic arm would have to constantly fashion a new administrative language to demarcate a moving landscape and new techniques for draining and fixing it. As we will see in the following chapters, parsing out land from water was both an engineering feat and a legal machination.

[65] Drawing on Noel Castree's classic reading about the capitalization of nature, scholars show how indeterminate ecological landscapes like fens, bogs and mires, to name a few, have been made into a social and capitalist space of production. Tom Mels, "Primitive Accumulation and the Production of Abstract Space: Nineteenth-Century Mire Reclamation on Gotland," *Antipode* 46, no. 4 (2014): 1113–33; Noel Castree, "Marxism and the Production of Nature," *Capital & Class* 24, no. 3 (2000): 5–36.

[66] It was not a matter of conceptual fixity, but the question of ownership in water was also not settled during this period.

[67] Petition of Lacam, HC 1806, Appendix 2, 26.

The problem did not simply end there. If the definition of where the land began and where the water ended was a problem, then the other geophysical challenge came from the flux in the measurement of the depth of water, a critical point affecting the navigability of the Company's cargo ships and army fleets. James Rennell declared Lacam's sounding of the water depth invalid. Lacam claimed that one of the reasons why his New Harbour was better suited for the task was that it boasted a more constant depth of water, thus providing safer passage than the older track, and was accordingly better suited for mooring for repairs than any of the existing harbors at that point. Lacam used the conventional lead and line to fathom the depth of the water, even though early forms of chronometers and sounding instruments had already been introduced.[68] Moreover, Lacam claimed experiential veracity over Rennell's scientific abstractions.

The courts fought a long and bitter battle starting in 1777 over the depth of the water, which failed to yield constant measures. As it turns out, marine soundings in the legal records were not innocent evidence, and entered the courts entangled in debates around law and insurance, thus muddying the boundaries between marine science and fiduciary concerns. By the time the case reached London, the questions that the controversy surrounding Channel Creek opened up were as follows: Can the evidence of "geographical science" be trusted? And if science was not corrupt should the judges then argue the case as a matter of individual accountability and honesty? All of this had to be settled in order to secure ownership in land–water admixture.

MEASUREMENT IN A CORRUPT WORLD

Human accountability is written into the history of instrumentation.[69] The law of evidence and standards played an important part in this history. However, what happens when measurements, or the failure thereof, are written into the history of property and ownership? In order

[68] Peter de Clercq, ed., *Nineteenth-Century Scientific Instruments and Their Makers* (Leiden: Museum Boerhaave, 1985), see especially Jim Bennett's chapter on John Harrison's unsuccessful attempts to refine and popularize his chronometer. "Instrument Makers and the 'Decline of Science in England': The Effect of Institutional Change on the Elite Makers of the Early Nineteenth Century," 13–28.

[69] James Poskett, "Sounding in Silence: Men, Machines and the Changing Environment of Naval Discipline, 1796–1815," *Journal of the British History of Science* 48, no. 2 (2015): 213–32.

to begin to deal with these questions, let us turn to the unfolding of the case and the subsequent arguments and debates that restructured the modalities of owning a mobile delta.[70] The history of European naval technology remains deeply enmeshed with the "discovery" of geographic novelties in the colonial environments. This intertwined history of technology and environment forces us to pay careful attention to the materiality of land and waterscapes.[71] This involves recognizing the two-step cognitive process that constituted Lacam's discovery of new geographical features.[72] The first step involved parsing out a moment in the fluvial timeline when deltaic mobility was named a permanent harbor. Lorraine Daston calls such naming "anchor[ing] permanent words to impermanent things, once and for all."[73] This was a classificatory act that condensed and flattened the variety and variability embedded within the fluid ecology of the landscape. The second step involved gathering scientific data through standards, measurements and customs to develop what I am going to call a scientific or "geographical bracketing of that space." Lacam demarcated, surveyed and named a particular land–water formation as the "New Harbour." In the act of calling this space a harbor, Lacam condensed the variety embedded in this marshy mobile landscape, but also attempted to flatten its temporal variability.

Describing the scientific project of developing a global language of cloud terminology, Daston pointed to the "collectively willed ontology" of seeing, parsing and naming nature.[74] In Lacam's case we get a glimpse of that same project of willed ontology, which took a different trajectory. This was not just an act of naming, but also the production of a collective act of seeing this space as a harbor. Daston's work on how a meteorological understanding of clouds was produced through the nineteenth century alerts us to this process. According to her, this particular "achievement of collective seeing and naming was made possible by terse descriptions that focused attention on a few key details and – even more important – obscured myriad of others. All classification depends on some degrees of abstraction from the blooming, buzzing world of particulars, accentuating some significant features and muting others."[75] What

[70] Steinberg, *Slide Mountain*, 3–21.

[71] John Law, "On the Social Explanation of Technical Change: The Case of the Portuguese Maritime Expansion," *Technology and Culture* 28 (1987): 227–52, 236.

[72] Here I am drawing on Lorraine Daston, "Cloud Physiognomy," *Representations* 135, no. 1 (2016): 45–71.

[73] Daston, "Cloud Physiognomy, 48.

[74] Daston, "Cloud Physiognomy," 53.

[75] Daston, "Cloud Physiognomy," 48.

remains specific to naming geographical formations, like their meteorological counterparts, is that they must be both temporally and spatially fixed before they change forms and enter a different taxonomic register.[76]

Yet, unlike Daston's project, Lacam's act of naming failed to become a collective venture. The language, metaphors and frames that Lacam and his supporters developed failed to globalize. After all, the Bengal Delta is a mobile landscape; what is silt today is swamp tomorrow, and perhaps headed for the bottom of the Bay of Bengal in the future. By developing, however indeterminately, a system of measuring the depth of the water to account for this space's practicability as a site for a harbor, Lacam attempted but failed to forge a common geographical consensus. This is why his case was an infrastructural failure, and in the process created legal dilemmas over the question of ownership of watery spaces. It was this act of collapsing variety and variability of this rich ecological world that would animate the discussions around the navigational metrics and legal terminology of accountability as Lacam's legal case began unfolding.

Following Lacam's petition against the Company, the judges in Calcutta convened a committee in 1783 to hear the oral evidence and look into the papers about this case that, in its final form, runs into over 750 handwritten and printed pages of material. On consulting the material, the committee divided the evidence into three categories in order to extricate the nature of geographical design and human error since it was difficult to ascertain whether Lacam was lying about the water depth or whether the other surveys were untrustworthy. The first two sections of evidence were geographical: one concerning the dangers of the older channel; and the other concerning the advantages of Lacam's proposal. The third section dealt with the human category. This consisted of papers documenting how and through what means Lacam acquired the tract of land from the government in Bengal, and on what basis his access to those grounds was later canceled.[77]

The judges also collected oral and written evidence from the merchants and mariners who traded and sailed in the Bay of Bengal, and catalogued them under the first two evidentiary categories to account for the relative military advantages of Lacam's New Harbour over the old channel.

[76] Daston shows that such a Linnaean system of naming was developed for species that evolve over generations, while marshes and land–water admixtures transform seasonally, annually and over a decade into something completely different. "Cloud Physiognomy," 49.

[77] Petition of Lacam, HC 1806, Appendix A (1), 7–8.

Under their last body of evidence, they turned to the depositions, requests and surveys submitted by Lacam. Lacam's proposal for the initial land grant was carefully crafted to request assistance to establish a wharf in a "place so well-formed by nature."[78] After the first failed survey of the area, undertaken by the Company, John Ritchie's 1776 survey approved the space as suitable for a harbor, although he refrained from commenting on the political expediency or economic viability for a harbor. Based on Ritchie's report, the Board of Directors and the Revenue Committee in Calcutta approved Lacam's land grant. While this land grant was meant to include both the landmass and the water bodies within the purview of the grant, Ritchie had his reservations. Although Ritchie had not conducted a formal cadastral survey of the area before, he was nonetheless deeply familiar with the nautical landscape of the Bay of Bengal and Hooghly River, having undertaken an extensive hydrographical survey of the rivers and coasts of the Bay of Bengal between 1767 and 1770.[79] Thus, he was familiar with the mutability of land and water in this landscape, even though he might not have had the terminology to explain the geological nature of the tidal delta and the changes it wrought upon the earth's surface. Ritchie's long familiarity with the landscape had also made him sensitive to the gap between the epistemological space of a land grant and the physical materiality of that space. As a legal and financial instrument, land grants of the delta fell short of representing the fluvial space in question.

Once the harbor was near completion, Lacam wanted a stipulation entitling him to the harbor and the profits arising out of this venture. More than demanding a privatization of this landscape, he wanted unconditional entitlement to tolls or taxes from the ships passing into his channel. This, he argued, was due to him for the gains the Company would achieve from his discovery of this new track to Bengal. He was particularly set on securing the toll he would receive from the passing ships and from any ship in need of his harbor for refitting. He also claimed that this would defray the costs of building the harbor. There were three critical voices of concern that stalled his project permanently. One of them came from Captain Ritchie, who doubted the political need for such a harbor in this mobile landscape. The second one came from Major Rennell, who raised doubts about Lacam's measurement.

[78] Petition of Lacam, HC 1806, Appendix 2, 19.

[79] John Ritchie, *An Hydrographical Journal of a Cursory Survey of the Coasts and Islands in the Bay of Bengal* (London: Ballantine & Law, 1784).

Finally, the third and perhaps the most powerful voice was that of Lord Cornwallis, the then Governor-General of Bengal, who doubted Lacam's personal integrity due to what he took to be the dubious nature of his profit motives for this venture. Swamps, it seems, produced murky characters who appeared in official records as political threats.[80]

In April 1790, the Committee met to decide the case of Benjamin Lacam. They began by consulting the original grant in order to assess the rights granted to Lacam and the equitable authority of the government to revoke it. Lord Cornwallis, who was then presiding over the Committee, finally decided that the passage into Lacam's new channel was "attended with the most imminent danger [and] consequently pronounced the scheme to be visionary and ill founded, and so far from being likely to yield affluence and credit to Mr. Lacam, or those concerned with him, the prosecution of it, I conceive, would more probably have entailed ruin and bankruptcy upon them." Cornwallis further added that Lacam's persistence in his plan, even after seeing the surveys denouncing it, "is further proof how much men engaged in schemes of speculative advantage may be led astray, how liable they are to have their judgment warped in favour of plans which their own zeal have projected."[81] With that the Company decided that it was right to revoke his land grant and dismissed his petition, but also offered him a handsome annuity of 600 pounds in compensation. After this decision, Lacam returned to England and attempted to settle his debts rather unsuccessfully. He barely escaped the debtor's prison. In 1803, Lacam made another petition to the House of Commons in London, after hearing that both the Company's and His Majesty's ships had ventured into his channel. After careful consideration of the written and oral records, the House again dismissed his petition, but not without raising his annuity to 1,000 pounds.

After that Lacam disappears from the archive, only to surface later in a vitriolic undated handwritten note probably from the first decade of the nineteenth century by a certain F. Barlow, who accuses him of selectively borrowing from Captain Ritchie's map by claiming it as his own. Barlow conjectures that Lacam did foresee that his track would be rendered

80 Nurfadzilah Yahaya's work on colonial Java explores swamps as sites of political distrust and legal vacuum: "How Kings Became Pirates: East India Company Maritime Jurisdiction," paper presented on December 12, 2016, at the Ocean of Law Conference, Leiden University. Daniel O. Sayers also explored this relation in *A Desolate Place for a Defiant People: The Archaeology of Maroons, Indigenous Americans, and Enslaved Laborers in the Great Dismal Swamp* (Gainesville, FL: University Press of Florida, 2014).

81 Petition of Lacam, HC 1806, Appendix H, 58.

unusable as a harbor, but adds, "It is of little importance now however – for it is most improbable that this track will ever be received under any circumstances short of a cataclysm."[82] New Harbour disappeared into the river just as Lacam vanished from history.

FLUID LANDSCAPES

Looking back at these records, it is surprising why the mobility of the landscape did not feature in the discussions about the factuality of the water depth, the search for accuracy within the instruments and the endless and nerve-wracking exchange over the character of Lacam. One probable explanation was the absence of any cognitive frames for understanding and capturing this tidal mobile landscape. Geological or earth sciences had yet to produce theories or narrative frames to explain the rapid flux that defined a deltaic landscape. The slow-moving changes wrought by the aqueous and igneous elements that created the earth's crust as a gradual process of accumulation and erosion, did not enter the popular discourses until the publication of Charles Lyell's *Principles of Geology* in 1830.[83]

The lawyers, sailors, hydrographers and bureaucrats involved in the case did not know how to explain the discrepancies of depth measurement outside the frames of metrics or the lens of human integrity and morality. It is difficult to speculate whether they were inhibited by bureaucratic constraints of codification or scientific terminological limitations. What we can deduce from the available sources is that this mutability and flux became pivotal for Lacam's venture. For Lacam, Channel Creek and its shores were a naturally forming harbor, and for his detractors it was a "place entirely of his own making." Both statements are perhaps true depending on one's interpretive frames. Thus, the Company, the Mayor's Court and later the House of Commons could easily nullify his claims of ownership to that place.

It is astounding that metrics and character integrity were the only two possible ways to debate the situation in New Harbour, particularly since the vernacular literary and folkloric landscape is rich with riverine

[82] Barlow Private Papers.

[83] William Whewell had offered a natural history theory of uniformitarianism as an answer to catastrophism within the scientific circles in the late eighteenth century. However, many of those ideas gained wider circulation with the publication of Charles Sir Lyell, *Principles of Geology* (London: John Murray, 1834).

metaphors that attest to the sand and silt that often overtake the river. These sources also reveal the nature of the water that eddies down to join two shores at once during the monsoon. Boatmen in Bengal live both on and off these half-soil, half-water areas. The English geographical terms "erosion" and "deposition" fail to capture the range of intimate river–human experiences that are contained in Bengali.[84]

In Bengali, for example, there are multiple ways of understanding erosion, signaling the diverse experiences of living with the river. *Chapa bhanga* is used when the river breaks away chunks of lands up to two meters wide as it flows during rainy seasons. *Bhanga*, on the other hand, refers to a form of erosion that wipes away a couple of acres of land in a matter of minutes and is experienced as a devastation that the monsoon brings regularly to this region. *Hanria bhanga* signals the strong undercurrents that sweep the soft, sandy bottom layer, causing the bank to hang precariously. Besides these three major terms, people of the lower Ganga basin employ other terms to understand the river's behavior: *Chechra bhanga* occurs mainly when floodwaters recede, revealing a newly emerged char; *bhurbhuri bhanga* captures the bubbling of the river eating away sandy lands; *nishi bhanga* describes nocturnal erosion; and *probol bhanga*, a superlative of *bhanga*, refers to devastating bank erosion.[85] In vernacular geological understandings water and land shared a polyvalent relation, each moving and shaping the other slowly and catastrophically. These theories existed much before the nineteenth-century British men of science, such as William Whewell, James Hutton or Llyell, came along with their uniformitarian analysis of the earth's crust.

Admittedly, it is difficult to translate and perhaps communicate in writing a language seeped in this watery vocabulary. Yet, it is precisely this wateriness that remains attuned to the mobility of land, water and everything in between the bounded space of land and the contained sphere of water. Contingency and indeterminacy are encapsulated not only at the rim where land and water mix, but also in our languages and our knowledge systems. It is these structures of knowing that shape and

[84] Hugh Raffles points to the same impossibility of translating the rich riverine language and the difficulties of communicating in writing a language seeped in this watery vocabulary, *In Amazonia: A Natural History* (Princeton, NJ: Princeton University Press, 2002), 27.

[85] This description is based on the work of Lahiri-Dutt and Samanta, *Dancing with the River*, 40–41; for more, see C. R. Abrar and S. Nurullah Azad, *Coping with Displacement: Riverbank Erosion in Northwest Bangladesh* (Dhaka: Rangpur Dinajpur Rural Service and North Bengal Institute, 2003), 17–18.

are shaped by our sensitivity to the fate of place in the larger discourses of space.[86] For the merchants of the East India Company, the changing depth of the tides appeared as an anomaly that could be fixed through either better instruments of measurement or the law. In the case of the New Harbour, such an anomaly led to an intense discussion around economic measures and legal instruments that might prevent further legal disputes of this kind. Ownership of such lands with water content exposed the limits of land grants. Unlike traditional histories around the contested nature of occupation and property in the colony, it was not a broken treaty but, rather, geography that exposed the limits of European epistemologies of property, which nonetheless failed to thwart the onward march of colonial capital. While the variability in the landscape needed fortification and strict methods of computing depth and tidal measurement, as the British would realize over the next century, only law could tackle this landscape to colonial advantage.[87]

Where are the voices of the indigenous people who have been living with the land, water, fish and tigers for centuries? If this chapter has detailed the negotiation between a British merchant and the Company around competing notions of entitlement and how ownership questions were inflected by existing ideas about fluvial landscapes, the science of measurement and geography, then what can we say about the indigenous notions of entitlement to those watery spaces? Although the Indians who were living there were surely involved in the construction work in the harbor, they remain invisible in the archive. The only time we have any intimation about what the indigenous population thought about Lacam's venture is when we sense the texture of human–land–water interactions that the legal and infrastructural debates cannot capture. John Ritchie ends his deposition with this "small voice of history":

> It is curious enough to observe the ideas which the country people entertain of the sinking of the wharf in Channel Creek: "Mr. Lacam," say they, "having erecting a very tall flag staff, and hoisted colors upon it, within the districts of the Gangee Saub, without first making the proper oblations and etc. to that Deity, he in a rage

[86] Edward Casey, *Fate of Place: A Philosophical History* (Berkeley, CA: University of California Press, 1998), 77–107.

[87] It is no wonder that Christopher Hill, in a different context, talked about the same preoccupation about measuring rainfall and the confusion of the colonial officials when they found that averaging annual rainfall in colonial India was not as easy as in temperate England. Christopher V. Hill, *River of Sorrow: Environment and Social Control in Riparian North India, 1770–1994* (Ann Arbor, MI: Association for Asian Studies, 1997).

at once sent the whole work to the bottom of the river, as a chastisement for so great a piece of presumption."[88]

This small voice, filtered as it is through the colonial archive, raises significant questions about ownership in a mobile world. "Gangee Saub" could only be a reference to the Hooghly River, a major distributary of the Ganga River . The locals believe that this river has the highest claim on the land that it creates since it is the river that gives them the land in the first place. Therefore, absolute ownership as envisioned by Lacam and the English law remained beyond the realm of the "natural" and all that is outside the purview of law and technology, namely, the natural movement of water and the silt it carries with it. The message of Ritchie's account by the unnamed "country people" is that the forces and energies of the river challenge the idea of the ownership of "landed" property, since what is land today through drying was yesterday's silt carried down by the river, and is now on its way to be deposited on the bed of the sea tomorrow. Riverine spirits acted as an ethical limit to human design and greed in the tidal delta.[89] They seem to be telling Lacam that land in this region is not atemporal and that one must learn to live with the temporariness of land and its flows and with water and its stagnations. Even if law abstracts silt into allodial rights, nature can always subvert it.

What transpired in this case was not merely a speculative venture gone awry, even though the colonial (and metropolitan) legal system could not locate who was responsible. Lacam's annuity (or compensation) secured legitimacy in the court's ruling against a harbor that had already begun to disappear into the river. One might argue that Lacam's New Harbour could not be rendered intelligible through the available representational practices. Thus, the members of Calcutta's Revenue Committee debated on how exactly to account for *water bodies* within the land grant made to Lacam. Others debated the limits of scientific instrumentation as evidence in a court of law. We can conclude that Lacam's frustration in the harbor-building speculation was not due to the lack of artifice or the failure of the (ill)-reputed "British perseverance and enterprise," but rather because the land in question was not ontologically available as land. Instead, the land had to be materially and conceptually produced within the legal and infrastructural ambits of the Company. As we will see in Chapters 2 and 3, such a spatial production occurred in three

88 Petition of Lacam, HC 1806, Appendix 2, 26.

89 Kalyanakrishnan Sivaramakrishnan, "Ethics of Nature in Indian Environmental History," *Modern Asian Studies* 49, no. 4 (2015): 1261–1310; Jalais, *Forest of Tigers*, 8–9.

interweaving registers: nature, custom or law, and science. The history of this space, narrated through revenue accounts and colonial legal debates, is ultimately a narrative about how the land formations were manufactured. They were manufactured through law and technology involving multiple failures in producing that space.

Even if Lacam's venture was a failure, though there might be some debate about the nature of such failures, it laid the groundwork for a particular arrangement of the space as an infrastructural landscape in this area and restaged the relation between law, geography and evidence. What was this particular arrangement? The following chapters turn to what I am calling the "manufacturing of a hydrological landscape." This manufacturing did not merely entail consolidation of the land and waterscape as various forms of privatized landholding. It also involved abstraction of various judicial and legal fictions of ownership, as well as of things "apparently inaccessible to private appropriation (private property): nature, the earth, life energies, desires and needs."[90] In the following two chapters I turn to those spaces within the infrastructural refashioning and hydroengineering of the landscape that resisted translations into mere economization and capitalist abstractions. In the process, I unearth those moments of empire-building where the ecological and sacral had to be accommodated into the envirotechnical reconstitution of these hybrid and aqueous spaces.

[90] Henri Lefevbre, *The Production of Space* (Oxford: Blackwell, 1991), 350.

2

Drying the Delta

Landholding in early nineteenth-century Bengal was not an isolated economic fact, but was instead bound up in an array of complex legal and ecological indeterminacies. If in the previous chapter Lacam's construction was threatened with disappearance, here we turn to what happened following the appearance of new land. While the presence of water in Lacam's landholding appeared as an anomaly, by the turn of the nineteenth century the Company officials in Calcutta began to create legal and engineering schemes to manage the swamps and thereby create landed property without water. This chapter turns to the legal processes through which the mobility of the landscape was accommodated into the architecture of ownership. The soaking ecologies of this swampy settlement became a site for productive law-making by creating open-ended possibilities within law for taking land. This chapter documents the haphazard extension and enactment of particular legal doctrines surrounding riverine land formations following the cancellation of Lacam's grant in the face of a landscape that refused to remain fixed.

In spite of the legal straitjacketing of the rich ecological world, closer attention to the distinct history of the marshes, alluvion and riverine deposition opens a new space for illuminating how the colonial property regime operated and highlighting the early exercises of quasi-state domain practiced by the emerging Company-State.[1] In the process the Company

[1] While the term "Company-State" was first used by Edmund Burke when he pilloried the East India Company for being a "state in the guise of a merchant," it was Philip J. Stern who deployed this as a concept to explain the hyphenated and hybrid nature of the

transformed itself from a trading body into a public agent managing land they arbitrarily designated for the general welfare. The Company's transformation was premised on using the new land formations in this estuarine delta to gradually institute a legal doctrine regulating alluvion by developing administrative categories. These categories became powerful tools to discipline the existing landed property relations in Calcutta through a regime of documentation. The mobility of the landscape in the Bengal Delta afforded the British the possibility of embedding illegitimacies into the legal architecture of landed property, which restructured the urban land markets in colonial India from the nineteenth century onward.

THINKING WITH WATER

Around 1804–5 the tidal Hooghly River started changing its course as it flowed to the Bay of Bengal. As a result, Calcutta, located at the edge of this tidal estuary, was exposed to the vagaries of the shifting river course. By the 1820s, the Hooghly had meandered so far west that it deposited alluvial land approximately three and a half miles long and roughly a quarter-mile wide along the western banks of Calcutta, opening up multiple ownership claims and new regulations on this strategic land–water admixture.[2] As we saw in Chapter 1, the movement of the river and the land formations was in flux in the Bengal Delta. Yet, for surveyors such as Ritchie, settlement officials and early historians, the riverine behavior in the tidal delta was considered anomalous even into the nineteenth century, and this anomaly had to be corrected and contained.[3] Describing the

sovereign power that the early East India Company employed. Philip J. Stern, *The Company-State: Corporate Sovereignty and the Early Modern Foundation of the British Empire in India* (Oxford: Oxford University Press, 2011), 4–5; for an engagement with the company as a military power, see Chris A. Bayly, "The British Military-Fiscal State and Indigenous Resistance: India 1720–1820," in *An Imperial State at War: Britain from 1689 to 1815*, ed. Lawrence Stone (London: Routledge, 1994), 322–54; Ian Bruce Watson, "Fortification and the 'Idea' of Force in Early English East India Company Relations with India," *Past and Present* 88 (1980): 70–87; as a trading empire, see Nigel Rigby, ed., *The Worlds of East India Company* (Woodbridge: Boydell Press, 2002).

2 Extract from the Proceedings of His Excellency the Most Nobel the Governor General in Council in the Territories Department, 24 March 1820, Judicial [Criminal], 1 April 1820, Prog. 15, §§1 and 2, West Bengal State Archive, Kolkata (hereafter WBSA).

3 This is not surprising given that the nineteenth-century engineers who surveyed the Hooghly River built their knowledge by working with surveyors whose expertise was temperate or Mediterranean rivers in Europe. A traditional knowledge circuit would begin at the Danube, followed by the Po, Vistula, Rhine and Adour, with their training

riverine behavior in the Gangetic plain, historian C. R. Wilson comments, "An Indian river in its old age is a thing full of caprice."[4] These capricious sedimentations became one of the most managed sites under the empire: managed through an array of legal denominations based on where and how they were formed.[5]

Riverine capriciousness was written into the legal history of land and fluvial ownership. The geological process of "dynamic sedimentation," which is a part of delta formation, created legal and infrastructural havoc across empires throughout the nineteenth century.[6] While not unique to the Bengal Delta, the process of drying the land by separating land from water played out differently in various parts of the world, with divergent impacts on the history of the landscape.[7] These sedimentations, sometimes adjacent to existing land, at other times linked with a fordable channel, and sometimes thrown up as vast islands arising from the bed of the river, became sites for technological and administrative experimentation in the empire. Yet, most of the existing literature on riverine cities focuses on the infrastructural projects of draining, which, although important in Bengal, were overridden by legal concerns. The colonial officials in Calcutta and the agrarian districts of Bengal understood that if riverine behavior vis-à-vis land formation and subsidence was not strategically managed, then these sedimentary spaces threatened to create a political vacuum for the Company. By turning to one such urban deposition along the western riverine limits of Calcutta, this chapter

traditionally ending with the rivers of England, primarily the Tyne. Irrigation Department Papers. Report on River Hooghly by H. Leonard, 1865, IOR/V/27/732/13 (British Library, UK).

4 Charles Robert Wilson, *Early Annals of the English in Bengal Being the Bengal Public Consultations for the First Half of the Eighteenth Century* (Calcutta: Thacker Spinck and Co., 1895), 103.

5 For a history of how the British managed such sedimentations within the agrarian context, see Nitin Sinha, "Fluvial Landscape," 209–37; Iqbal, *The Bengal Delta*. For a history of how massive irrigation infrastructure transformed a rain-fed landscape into flood control zone, see D'Souza, *Drowned and Dammed*. For an anthropological account of life in these transient environments, see Lahiri-Dutt and Samanta, *Dancing with the River*.

6 Charles R. Kolb and Jack R. Van Lopik, "Depositional Environments of the Mississippi River Deltaic Plain." In *Deltas in Their Geologic Framework*, ed. Martha Lou Shirley (Houston, TX: Houston Geologic Society, 1969), 17.

7 For studies exploring the colonial struggle of defining coastlines in other parts of the world, see Carter, *The Road to Botany Bay*; Kelman, *A River and Its City*; Booker, *Down by the Bay*; Vera S. Candiani, *Dreaming of Dry Land: Environmental Transformation in Colonial Mexico City* (Stanford, CA: Stanford University Press, 2014). This is not an exhaustive list but it signals that the struggle was not unique to the Bay of Bengal.

documents how ecological differences opened up a space for legal exceptionalism within property laws in the empire.

"More land is being yearly added," bemoaned a Company servant out of frustration at both mapping and estimating the costs of converting these marshy deposits into a recognizable riverbank.[8] The distance between the dotted line and the continuous line in Figure 2.1 shows the extent of deposition along the bank of the Hooghly River. The solid black line is the border between land and water as it existed in 1793, giving us a clear sense of the length and breadth of the built-up area in the subsequent century. While it is impossible to measure a moving landscape, we do know that the land kept growing, which the newly formed Territories Department under the aegis of the Justices of Peace designated as "shoals or mud flats."[9] The fickle river rendered the 1797 land and ground measurements in Calcutta obsolete.[10]

How was this perceived environmental "anomaly" managed through the nineteenth century? During this period, preoccupation with the specific ecology of the delta entered the domain of scientific curiosity and environmental authority, because it interrupted agrarian revenue flows and destroyed urban property and infrastructures. While the Bengal Army of engineers built slopes into the river in an attempt to fortify and reclaim the bank as "dried" land, drying proved to be beyond what engineering alone could achieve. This was not the first attempt at land reclamation under the British in India. In 1784, engineers in the Bombay Islands constructed an embankment around Worli

8 In a letter dated 26 September 1854 from C. T. Cockburn, the Chief Magistrate of Calcutta, to the Hodgson Pratt, the Secretary of Bengal, Cockburn complains about the difficulties and expenses involved in leveling and laying out the constantly shifting limits of the city. "In reply I beg to state that the Strand Bank in question is I think sufficiently consolidated for the purpose stated, but having been formed through a succession of years, some trouble and expense will be necessary in levelling & laying it out." Judicial [Judicial], 19 April 1855, Consultation Nos. 71–72, WBSA, Kolkata.

9 This parcel of land had many names, "shoals and mudflats" being just one of them. Extract from the Proceedings of His Excellency the Most Noble the Governor General in Council in the Territories Department, 24 March 1820, Judicial [Criminal], 1 April 1820, Prog. 15, WBSA, §71. After the survey, the Strand Bank Fund was set up and they named the land "shoals or mud flat." Letter from S. Wauchope, Commissioner of Police to the Secretary to the Government of Bengal, 27 April 1863, Strand Bank Funds, Judicial [Judicial], October 1863, Prog. 123–31, WBSA, Kolkata.

10 Extract of the Territories Department, 1 April 1820, Judicial [Criminal], Prog. 15, §§11, 13, 15, 16, 17 and 21, WBSA, Kolkata. This report refers to a report made by Mr. [Francis?] Gladwin in 1797, who prepared a ground measurement of the properties along the river frontage. It also mentions two other reports dated in the Bengali Calendar but the names of the surveyors are not mentioned. I could not locate any of these reports.

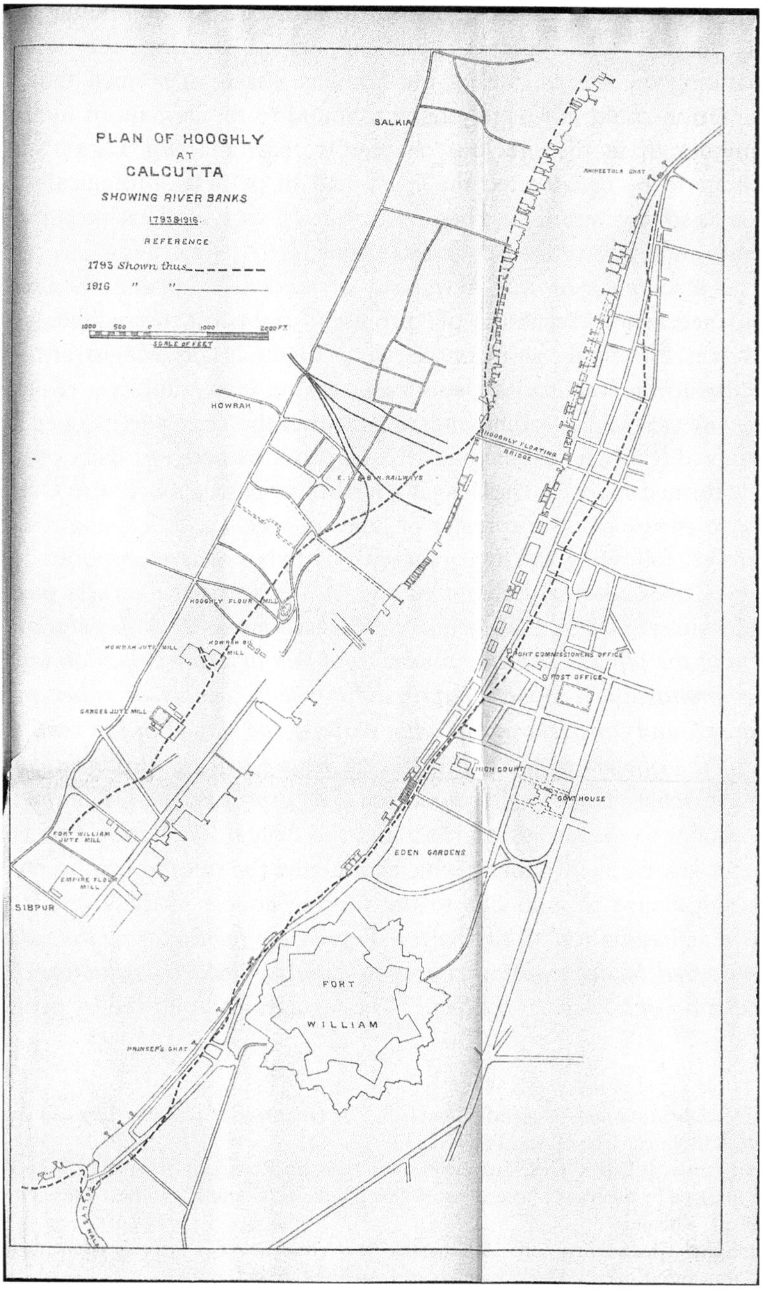

FIGURE 2.1 Fortified Strand Bank land from 1793 to 1916, *Report on the Hooghly River and Its Head-Waters: Vol. 2 Maps and Plates*. Calcutta, 1919, Map 17 © British Library Board IOR/V/27/732/15.

Creek that surrounded the settlement to protect it from flooding during high tides.[11] While that experiment helped develop the craft of embanking the creeks dotting the Bombay coasts, it seemed that land reclamation could not merely remain bound to the domain of hydraulic techniques. If in the previous chapter we saw that for Lacam's New Harbour to be constructed the space had to be first ontologically produced as solid ground, then here we witness how a legal architecture had to craft and endorse the act of reclaiming.

The emergence of this new strip of land as shown in Figure 2.1 broadened into a discussion of "property" rights on the sediments of a tidal river. Noting the shifts in the river, Calcutta's City Magistrate sent a directive to the Territories Department asking it to conduct a riverbank survey in 1820. The recommendations within the 1820 survey opened up protracted legal and municipal debates about ownership rights over the newly formed space.[12] The Bengal government passed a bylaw in October 1814 to ensure smooth transfer of land for roads and draining the city. However, following the 1820 survey this bylaw was incorporated into Bengal Codes, as Regulation 1 of 1824.[13] This was the earliest piece of ruling with regard to land acquisition, also known as *A Regulation for enabling the officers of Government to obtain at a fair valuation land or other immovable property required for roads, canals or other public purposes, and for declaring in what manner the claims of the zemindárs and of the officers in the Salt Department are to be adjusted in certain districts, where lands are required for the purposes of salt manufacture.* Subsequent to the survey of 1820 and the Bengal Regulation I of 1824, another law came into force, which stipulated the rules to be observed in assessing claims to land due to the shifting course of the river. It was known as Regulation XI of 1825: *A Regulation for declaring the rules to be observed in determining claims to Lands gained by Alluvion or by Dereliction of a River or the Sea.*[14] The regulations, followed by petitions

11 S. M. Edwardes and James M. Campbell, *The Gazetteer of Bombay City and Island*, vol. 2 (Bombay: Time Press, 1909), 120–21.

12 Magistrate to Direct Their Surveyors to Survey and Report on the Bank of the River Mentioned by the Mint Master, Fort William, Judicial [Criminal], 25 June 1807, Prog. 1, WBSA, Kolkata.

13 Bidhisha Chakraborty and Sharmistha De, eds., *Select Documents on Calcutta 1800–1900* (Kolkata: Directorate of State Archives, 2012), 2.

14 Charles D. Field, ed., *The Regulations of the Bengal Code with a Chronological Table of Repeals and Amendments, an Introduction, Notes and an Index* (Calcutta: Thacker, Spink and Co., 1875).

and legal cases, made up the complex tapestry that secured legitimation for taking land and reordering the relations between the river, the city and its inhabitants.

Drying a tidal landscape was as much an infrastructural project as it was an ontological endeavor in producing a dry culture with colonial law as its handmaiden. While infrastructural projects involving drainage and fortification attempted to dry the land, members of the Territories Department turned to juridical procedures. Through these procedures, the marshes came to occupy multiple legal denominations ranging from "accretions" and "derelicted land" to "disappeared shoals," setting the stage for their gradual transformation into landed property. While legal denominations slowly crafted these watery soils into property in Britain's eastern colony, the process was not without fierce contestations. Bit by bit, these watery terrains of the "shoals or mud flats" entered political modernity as "acquisitionable" property. Hydrological infrastructures working with the force of law began to materially concretize conquest in Britain's eastern colony from the late eighteenth century on.[15] This colonial struggle of converting marshes into urban property thus, provides an excellent lens into how British bureaucratic rationality managed geographical indeterminacy.[16]

This chapter argues that the infrastructural battles with water that were often overwhelmed by the tides of the Hooghly must be read not merely as the failure of colonial engineering, but also as productive moments in the making of a particular kind of property in the marshes. The legal separation of land from water in the nineteenth century enabled the emergence of a particular juridical logic governing urban property and state authority in the creation of "acquisitionable" land. In the process, titles also circulated, as British "squatters" became landlords through the fabrication of legal fictions and the deployment of open-ended potentialities within the laws of occupation in the colonies. The urban story of land acquisition is thus another aspect of the Hobbesian form of sovereignty that the Company-State acquired over Bengal by force.

A malleable juridical reasoning eased the process of land-grabbing by the Company under the guise of extensive infrastructural projects as force

[15] On the different ideologies driving European conquest and expansion in the Atlantic and Asia, see Andrew Fitzmaurice, *Sovereignty, Property and Empire, 1500–2000* (Cambridge: Cambridge University Press, 2014).

[16] Benton, *A Search for Sovereignty*, 22–23.

was turned into legal authority and shaped the Company's relation to the urban landscape. Studying the construction of a riverbank in Calcutta therefore, helps us understand what animated the ideologies driving expansion of public works, and the legal logic of securing the newly formed spaces as Company lands from the early decades of the nineteenth century. Many of the major urban infrastructural projects were made possible through emergency provisions to acquire so-called indeterminate landscapes, which were neither bounded land nor flowing water.[17] These provisions also characterize the Company's modes of bureaucratic conquest over plantations, lands and markets. I term these unstable bodies of legal encoding that enabled such conquests the "quasi-state domain practices" of the Company-State.

HYDROLOGICAL MOMENTS IN COLONIAL HISTORY OF PROPERTY

The logic of water shaped the laws of land and property through the nineteenth century. In order to understand that claim, let us briefly turn to the history of water and sedimentary deposits in British India. I am terming this era of fixing perceived geographical indeterminacy through law the "hydrological moment" in the history of property in colonial India. As scholars working on South Asia and elsewhere have argued, managing water and the movements of water bodies is deeply entangled in projects of state-making.[18] Colonial hydraulics was one of many such projects.[19] The various imperial regimes of water manifested themselves primarily through the canalization of rivers from

[17] Hussain, *The Jurisprudence of Emergency*, 6–7.

[18] For a seminal and much debated early argument, see K. A. Wittfogel, "The Hydraulic Civilizations," in *Man's Role in the Changing Face of the Earth*, ed. W. L. Thomas (Chicago, IL: Chicago University Press, 1956), 152–64; for works that expand and critique Wittfogel, see Donald Worster, *Rivers of Empire: Water, Aridity, and the Growth of the American West* (Oxford: Oxford University Press, 1982), and Denis Cosgrove and Geoff Petts, *Water, Engineering and Landscape: Water Control and Landscape Transformation in the Modern Period* (London: Bellhaven Press, 1990); for a more recent work on technological remaking and state-building, see Hill, *River of Sorrow*; David Mosse, *The Rule of Water: Statecraft, Ecology and Collective Action in South India* (New Delhi: Oxford University Press, 2003); D'Souza, *Drowned and Dammed*; Pritchard, *Confluence*, explored similar concerns in the French context.

[19] The management of deltas and estuaries and the management of the Himalayan watershed reveal two distinct and contrasting histories. For a contrast to the deltaic narrative, see Daniel Haines, *Building the Empire, Building the Nation: Development, Legitimacy and Hydro-Politics in Sind, 1919–1969* (Karachi: Oxford University Press, 2013).

the middle decades of the nineteenth century. In the course of this period, the hydraulic paradigm in colonial South Asia transitioned from one when centralized systems of colonial irrigation began to slowly replace indigenous forms of water management.[20] A cursory glance at the professionalization of engineering in colonial India starting from 1848, shows how engineering was directly linked to the growing demand for irrigation technologies.[21] David Gilmartin's work demonstrates how water engineering transformed the relation between nature and productivity. In this transformation, waste emerged as an engineering and economic category. He further argues, especially in his study of the Doab in the Punjab region, that the management of water also transformed communities beginning from the second half of the nineteenth century.[22]

Transitions in systems of managing water etched the fault lines of community formations across the Indus- and the Ganga-Brahmaputra river valleys.[23] Irrigation infrastructure redrew the riverine courses, fixed them through "improved" embankments and restaged water through new epistemologies of waste and value.[24] As irrigated water became an

See also Ken Pomeranz, "The Great Himalayan Watershed: Agrarian Crisis, Mega-Dams and the Environment," *New Left Review* 58 (2009): 5–39.

20 Building upon Wittfogel's contested work on hydraulic civilization, a range of scholars have turned to water to study its political economy, investigating how water and its infrastructures are embedded within questions of power, sovereignty and techno-science. D'Souza's work makes a robust argument for understanding the embedded nature of water and governance as a paradigm for understanding shifts in state power, production of knowledge and controlling space and people. See D'Souza, "From Conflict to Securitizing Water: The Making of Another Hydraulic Paradigm in India?," *Water Security in South Asia, Peace Research Institute Oslo* (Norway), 14–15 June 2011, www.prio.org/Global/upload/CSCW/PrioRohan.doc, accessed November 2015.

21 Arun Kumar, "Colonial Requirements and Engineering Education: The Public Works Department," in Roy McLeod and Deepak Kumar, eds., *Technology and the Raj: Wester Technology and Technical Transfers to India, 1700–1947* (Sage Publications: Bew Delhi, 1995); see also K. V. Mittal, *History of the Thomason College of Engineering 1847–1949* (Roorke: University of Roorke, 1986).

22 Gilmartin, *Blood and Water* and "Scientific Empire and Imperial Science: Colonialism and Irrigation Technology in the Indus Basin," *Journal of Asian Studies* 53, no. 4 (1994): 1127–48.

23 Gilmartin, *Blood and Water*, and Iqbal, *The Bengal Delta*, make a convincing case for rethinking peasant politics and community formation in the Indus and Gangetic river basins, respectively, from the perspective of colonial techno-scientific water control.

24 Elizabeth Whitcombe, *Agrarian Conditions in Northern India: The United Provinces under British Rule 1860–1900*, vol. 1 (Berkeley, CA: University of California Press,

agrarian commodity within the changing political economy of the nineteenth-century world, traditional knowledge about using ground and rainwater, as well as vernacular epistemologies of land–water relations, was gradually rendered obsolete.[25]

Nonetheless, water's role in history was not only as a commodity, capitalist or otherwise. As the case of the Bengal Delta testifies, there were many geographical formations between land and water that often remained beyond the grasp of the early British surveyors.[26] The British made both successful and unsuccessful attempts at drying the shoals, mudflats, marshes, char, diara and even the land between high and low tides. In most instances, as is wont to happen in tidal or active deltas, such soaking ecological formations became "frequent sources of contention and affray," piling up innumerable cases in the Court of Sadr Diwani Adalat (district courts) and the Privy Council.[27] In 1825, the Court of Justice in Calcutta gathered legal authorities on existing customs from Hindu scriptures and Islamic law, as well as precedence in cases settled during the last quarter of the eighteenth century to arrive at a legal doctrine about settling disputes pertaining to alluvions.[28] Through this ruling, the ecological variety of this tidal delta was finally subsumed into three distinct legal categories: (1) gradually accreted land, (2) sudden emergence of alluvion and (3) derelicted land, or disappeared shoals within riparian laws.[29] These distinctions became central not only to the British as they

1972); Ian Stone, *Canal Irrigation in British India: Perspectives on Technological Change in a Peasant Economy* (Cambridge: Cambridge University Press, 1985); Imran Ali, *The Punjab under Imperialism, 1885–1947* (Oxford: Oxford University Press, 1987); M. Mufakharul Islam, *Irrigation, Agriculture and the Raj: Punjab, 1887–1947* (New Delhi: Manohar, 1997).

25 Anil Agarwal and Sunita Narain, eds., *Dying Wisdom: Rise, Fall and Potential of India's Traditional Water Harvesting Systems* (New Delhi: Centre for Science and Environment, 1997); Stone, *Canal Irrigation.*

26 For a discussion on how the geographical knowledge based on temperate climatic system failed to translate in colonial India, see Hill, "Introduction: The Environmental Laboratory," 1–9.

27 Field, *Bengal Code*, 580.

28 Cases used in the 1825 ruling as precedence are as follows: *Issurchand Rai and Others v. Ramchand Mukherji* (Sadr Diwani Adalat S.D.A., 221; 11 December 1807), *Raja Grieschand v. Maharaja Tezchand* (S.D.A., 274, n.d.), *Radhamohan Rai and Others v. Suranjan Banerji* (S.D.A., 319), *Kunwur Harri Nath Rai v. Mussamat Jyedurga Barwan* (S.D.A. 269), *Zibu Nia v. Persan Rai* (S.D.A. 316) and *Ramkishen Rai v. Gopi Mohan Babu* (S.D.A. 340). Cited in Field, *Bengal Code*, 580–82.

29 Lal Mohun Doss, *The Law of Riparian Rights, Alluvion and Fishery: With Introductory Lectures on the Rights of Littoral States over the Open Sea, Territorial Waters, Bays,*

sought to manage the landscape, but also to demarcate the Company-State's ownership in newly formed lands and riverine accretions.

Water occupied a very particular currency within this legal architecture that went beyond sharing, embanking and delineating rivers and seas into distinct legal units. Water in the colonial context defined many of the laws of landed property. This was not just the case in the colonies, for water in fact straddled multiple codes and legal doctrines even in England. Water law is one of the most "Romanesque" aspects of English law.[30] As Joshua Geztler's illuminating study reveals, an unstable body of laws and legal doctrine in water began to emerge in the British common-law courts from 1750. These new water doctrines were based on older concepts of common goods, but also incorporated natural law codes of Roman and civilian law.[31] It is no wonder then that laws governing alluvion ownership in England borrowed heavily from Roman laws.

Encounter with the unstable colonial geography and the developments in the English courts began to shape the haphazard extension of laws in flowing water, which transformed land laws across the empire. In the drafting of Bengal Regulation XI of 1825, for instance, the Indian judges who were invited to comment on the existing practices around alluvion, pointed out that fluvial borders between estates were governed by the custom of *dhardhura*.[32] According to this custom, if a certain portion of land is suddenly moved to the other bank, and if the land was still recognizable after the shift, (for instance through physical features like trees or shrubs), then the previous owner might be able to make a claim to it.[33] However, the decisive factor in such cases was whether accession was gradual or sudden. In order to decide that, British officials sought commensurability of the dhardhura custom with Roman law, which defined alluvion as "an imperceptible increase ... which is added so gradually

&c., and the Rights of the Crown and the Littoral Proprietors Respectively over the Fore-Shore of the Sea, Tagore Law Lectures (Calcutta: Thacker, Spink and Co., 1891).

30 Joshua Getzler, *A History of Water Rights at Common Law* (Oxford: Oxford University Press, 2004), 117–52.

31 See especially his reading of how Bracton adopted Romanist and natural-rights conceptions of water law, which were never "nativized within the English common law, though their presence was always felt." Getzler, *A History of Water Rights*, 117.

32 *Nasiruddin Ahmed and Others v. Mussamat Umadi and Others* (IV. N. W. P. Rep. Civ. Ap. 1), *Dulkin Harpal Kunwari v. Ubruck Singh and Others* (IV. N. W. P. Rep. Civ. Ap. 18) and *Mussamal Rani Katiyani v. Sheik Mahomed Sharf-ud-din* (IV. N. W. P. Rep. Civ. Ap. 189) cited in Doss, *The Law of Riparian Rights*, 228–29.

33 Cases decided on the custom of dhardhura were often difficult to prove. Doss, *The Law of Riparian Rights*, 229.

that no one can perceive how much is added at any one moment of time."[34] Using the Roman principle of *accessio*, which designates both increase in landed surface and the mode of increase, the 1825 Regulation developed intricate distinctions between gradual accretion and sudden appearance of alluvion. Yet, many of these arguments often failed to account for the seasonal nature of land that would be submerged during annual floods, sometimes even for a few years, only to appear and then seasonally disappear.[35]

From the early nineteenth century, we begin to witness an intense scrutiny of geography and riverine behavior in the colonial courts, which began to define these ecological formations as a problem of ownership over swamps, marshes, wetlands and bogs. Scholars studying such hybrid landscapes in various parts of Asia argue that the ecological volatility caused these spaces to be treated as "resource frontiers."[36] In order to receive uninterrupted revenue from these new land formations, the Company-State enacted regulations through which they sought to abstract these spaces from their existing ecologies. These land–water admixtures were mostly alluvial sediments (also known as *puli mati*, char or diara) and were extremely fertile spaces. Thus it is no wonder that the various property regimes that inhered in these fluid geographies were both precarious and speculative.[37] Due to the shifting and changing nature of the landscape, such frontiers were often difficult to measure and harder to settle permanently even after the passing of the Permanent Settlement Act of 1793.[38] Unlike bounded land, these environmental formations continued to challenge both the revenue extraction and fixation of proprietorial rights through much of the nineteenth century.

[34] Field, *Bengal Code*, 584.

[35] Similar cases cropped up in other parts of the world. In her recent work on Iberian Empire, Tamar Herzog explores similar dilemmas around moving landscape. See "Moving Island in a Sea of Land: 1518–1864," in *Frontiers of Possession*, 191–242.

[36] I borrow this idea from Anna Lowenhaupt Tsing, *Friction: An Ethnography of Global Connections* (Princeton, NJ: Princeton University Press, 2005). Tsing primarily looks at the twentieth-century creation of disputed urban waterfronts in Asia, which is seen as a frontier space. See also David A. Biggs, *Quagmire: Nation-Building and Nature in the Mekong Delta* (Seattle, WA: University of Washington Press, 2010).

[37] Sinha, "Fluvial Landscape," 210–11.

[38] The literature on the search for fixity with regard to revenue and the 1793 Permanent Settlement of Bengal is huge and would be a diversion to go into here. However, it might be worthwhile to note Robert Travers's assertion that it was fixity, rather than an equitable tax structure, that guided much of the agrarian land settlement in Bengal. Travers, "'The Real Value of the Lands.'"

Such landscapes, deemed as "wastelands" in the agrarian revenue rolls, were placed under direct Company management.[39] Newly formed lands were also known as *benami* lands (literally, lands without names or titles), and such lands would enter the colonial regime of documentation when they had been in existence for twelve continuous years.[40] Often what that meant in practice was that an intermediary would be sent to manage these agrarian spaces on behalf of the Company. This would ensure that even if the farmland disappeared, the monetary loss would be borne by the intermediaries, who were mostly Indian speculators. This intermedial credit system then acted as a financial buffer securing for the Company uninterrupted tax flows by transferring the ecological risk to the Indian financiers.[41] However, such risk transfer did not work in the urban riverine landscapes. In the urban riverine landscapes, the disappearance of land would result in the built environment being washed away, as in the case of Lacam's harbor. Riverine movements affected built-up property, such as buildings, harbors and roads, very differently from agrarian lands. Loss of infrastructure could not be recovered by readjusting revenue extraction, by using financial buffers and intermediary structures that worked in protecting the British in instances of agrarian land subsidence.

Urban ownership of this partially aquatic terrain – a landscape that was neither bounded land nor flowing water, but land soaked in water – had to be managed differently. In a recent rethinking of the Company's status, Bhavani Raman's study of *mirasdari* (a form of landholding) in nineteenth-century Madras Presidency persuasively argues that the Company not merely was a tax collector, but also fashioned itself as a land developer.[42] Studying the processes through which the Company

[39] According to D'Souza, *Dammed and Drowned*, and Iqbal, *The Bengal Delta*, intermediary tax collectors controlled much of the revenue collection apparatus and credit and debt markets in these landscapes during the first decades of the nineteenth century, as the Company was in the process of devising new laws and embankment practices. These intermediaries acted as financial buffers to the perceived ecological indeterminacy, which could affect the extraction of revenue if the land disappeared or was flooded.

[40] Iqbal, *The Bengal Delta*, 53–56.

[41] For an account of how new laws of measurement were introduced to manage the constant accretion and dereliction of lands formations in the Gangetic plain, see Sinha, "Fluvial Landscape," 209–37.

[42] Bhavani Raman, "Sovereignty, Property and Colonization: The East India Company and Land Development in Madras," *Journal of Economic and Social History of the Orient*. Special Issue on "Repossessing Property in South Asia: Land, Rights and Law across Modern/Early Modern Divide" (forthcoming 2018).

intervened in the urban fluvial landscape to secure tenurial rights in marshes reveals how the Company calibrated its sovereignty as both a tax collector and land developer. As we turn to the next section, we will see that from the last decades of the eighteenth century on, multiple hydrological regimes began to proliferate, not just in the agrarian landscape, but also within the urban space through which the British merchants demarcated ecologically volatile landscapes into new forms of aquatic resource frontiers. These aqueous landscapes within urban space did not become title-less lands managed by intermediaries, but instead charted a new terrain in colonialism's government of paper.[43]

THE CURIOUS CASE OF THE LAND THAT GREW: HOW TO OWN CALCUTTA?

Paper, both nominally and in its materiality, was necessary for securing legitimacy for the Company's "conquest" of Bengal. Yet how did this play out in the urban context of Calcutta? Moreover, what role did the movement of the river in 1804–5 and the deposition of extra land play in securing conquest through regimes of documentation and an ordering of legal categories? The eighteenth century saw the beginning of what would be multiple attempts at consolidating land titles and holdings in Calcutta over a period of almost 150 years. Four years after the British established Calcutta, they began to acquire the surrounding terrain by "taking advantage of the *unsettled* state of Lower Bengal."[44] Attempts were repeatedly made to "systematically" document landholdings in Calcutta. But documents, much like land, were exposed to the vagaries of climate and animals. Zephania Holwell, the tax collector who assumed the office of

[43] Richard Saumarez Smith speaks of the unstable category of records, regulations and rules and the disjuncture between the empire of paper and the everyday lived realities of tax collection, "Rule-by-Records and Rule-by-Reports: Complementary Aspects of the British Imperial Rule of Law," *Contributions to Indian Sociology* 19, no. 1 (1985): 153–76. Matthew Hull's anthropological work reengages some of these concerns in his study of bureaucracy and the crafting of urban space, *Government of Paper: The Materiality of Bureaucracy in Pakistan* (Berkeley, CA: University of California Press, 2012), and for a historical account of the culture and politics of bureaucracy, see Bhavani Raman, *Document Raj: Writing and Scribes in Early Colonial South Asia* (Chicago, IL: University of Chicago Press, 2012).

[44] Reginald Sterndale, *An Historical Account of the Calcutta Collectorate, Collector's Cutcherry or Calcutra Pottah Office, from the Days of the Zemindars to the Present Time, with a Brief Notice of the Zemindars and Collectors of Calcutta, the Ground Tenure or Land Revenue System, Town Duties, Excise and Stamp Revenue* (Calcutta: Bengal Secretariat Press, 1885), 4 (emphasis mine).

the Calcutta Cutcherry (Collectorate Office) in 1752, attempted to trace the "accounts of the gunge [town] before it was farmed." However, he writes that "[he was] told that they were washed away in the great storm [in 1738]," and when he inquired for the period between 1738 to 1752, he was told "the white ants have destroyed [the records]."[45]

Ants and storms notwithstanding, he finally produced one of the earliest recorded accounts of the city's landholding and submitted it to the Court of Directors of the East India Company in London. The town of Calcutta then consisted of 5,472.5 bigha, from which the Company drew a ground rent of three rupees sicca per bigha per annum.[46] The Company had acquired only 310 bighas as its own land. Apart from this land, a considerable portion of land, amounting to 3,650 bighas, was scattered across the British settlement in Calcutta and its southern limits, which were not under the Company occupation. According to William Bolts's famous *Considerations on Indian Affairs* (1772), "independent proprietors" owned the 3,650 bighas of lands in Simleah, Molunga, Hogul Courea (Hogulkuria) and Mirzapore. These proprietors were deemed independent because they refused to acknowledge the authority of the Company to levy rent and refused to pay it for twenty-five years of the Company's presence in Calcutta, from 1690 to 1715 .[47]

Even though the Company had acquired revenue rights to Bengal through a royal decree from the Mughals, the lineaments of possessive authority over the ground, soil, and watery admixtures were highly contested.[48] John Surman, a merchant and one of the earliest tax collectors of Calcutta, failed in his attempts to purchase the land from these independent proprietors; therefore, he "wrested [those lands] from them by fraud or violence."[49] It may be interesting to note that by the time Reginald C. Sterndale, the official Company historian of the Calcutta Collectorate, quotes Bolts's *Consideration on Indian Affairs* in 1885, he changes the word "violence" to read "fraud and force."[50] Apart from glossing the nature of this force, Sterndale informs us that there were no

[45] Farm was used to denote leasing lands in Calcutta and farming in this context has nothing to do agriculture. Quoted in Sterndale, *Calcutta Collectorate*, 1.

[46] Sicca was a form of silver monetary unit used across the Gangetic plain. Sterndale, *Calcutta Collectorate*, 8.

[47] William Bolts, *Considerations on India Affairs, Particularly Respecting the Present State of Bengal and Its Dependencies* (London: J. Almon, 1772), 89–90.

[48] For more on the crisis and anxiety in establishing a new colonial order of agrarian land settlement, see Wilson, *Domination of Strangers*, 45–74.

[49] Bolts, *Considerations*, 229.

[50] Sterndale, *Calcutta Collectorate*, 8.

records left as to how the Company came to acquire those lands, but added that the Company received those so-called unsettled parts of the 24 Parganas by ousting the zamindars who lived there.[51]

From 1720 onward the Company made many partially successful attempts to collect ground rent in Calcutta and to record landholdings.[52] The earliest recorded lease in the Cutcherry stemmed from 1728, but a paper-based authority of titles and deeds did not always succeed in backing the property market, even as late as the mid-nineteenth century. Instead, land and property often remained caught in a world of credit, trust and nonmonetized tenancy relations, where even the nature of property evaluation was in constant flux.[53] While early attempts at collecting ground rent met with resistance, by 1758, directly following the Battle of Plassey, the British finally acquired the town of Calcutta as a fee-simple possession, i.e. as absolute ownership in freehold property, and eventually legitimized their title as the zamindars of Calcutta. Nonetheless, rent collection remained a fraught affair for some time to come.

The bureaucracy involved in collecting ground rent and registering landholding through *pattahs* (forms of leases) was a braided affair. The *Hobson-Jobson Glossary* defines the pattah as "the conditions on which

[51] Sterndale, *Calcutta Collectorate*, 9. Frederic Millet's minutes also testify to this. See Papers Regarding the Administration of Civil Justice in the Bengal Presidency – Compilation by Frederic Millett of a New Code of Civil Procedure, vol. 1 IOR/F/4/1762/69547: Oct. 1834 [June 1838], British Library, UK.

[52] Although the signature remains undecipherable, Sterndale conjectures it to be that of one Mr. Freke. For more details, see Sterndale, *Calcutta Collectorate*, 15–17.

[53] I am not denying that there was no system of demarcating property. What I am showing is the absence of the kind of centralized documenting and monetization the British were after. Before the British, the Mughal Empire had developed a complex system of agrarian property. For works on the Mughal and early modern documentary practices, see Frank Perlin, "Proto-Industrialization and Pre-Colonial South Asia," *Past and Present* 98, no. 1 (1983): 30–95. Scholars have argued that, while the British adopted much of the existing Mughal land tenurial system, by rendering land an object of private property the British nonetheless facilitated the transformation of existing concepts of property, sovereignty and value. For a discussion on the question of transformation of concepts, nature of power and value, see Bernard S. Cohn, "Representing Authority in Victorian India," in *The Invention of Tradition*, ed. Eric Hobsbawm and Terence Ranger (Cambridge: Cambridge University Press, 1983); Travers, "'The Real Value of Lands'"; for an overview on the nature and effects of colonial property regime, see Sturman, *The Government of Social Life in Colonial India*, 35–70; for accounts that document the vibrant operation of both credit and metallic currency, see Rajat Ray, "Asian Capital in the Age of European Domination: The Rise of the Bazaar, 1880–1914," *Modern Asian Studies* 29, no. 3 (1995): 449–554; Sen, *Empire of Free Trade*; Yang, *Bazaar India*.

lands are held; a lease or other document securing rights in land or house property."[54] To give an early example of a pattah:

A pottah being granted unto Lokicanto Seat for 6 cottahs and 8 chittacks in ground in bazar Calcutta, rent 15 annas 7 pies sicca per annum. Calcutta cutcherry, this 3rd day of January 1758, no. 1 [signed] M. Collet, Zemindar.[55]

While stating the amount of land owned by the proprietor and the form of ownership, the pattahs did not mention the boundaries of the land in question. An early nineteenth-century pattah should not be confused with a property title or deed. Indeed, the records of the pattah office were held not with the Collectorate office, but with the local police, and they were often appended to the town surveys.[56] Although it was not uncommon for some of the wealthy residents of Calcutta to have wills, a documentary regime backing the property market had not yet been standardized. For instance, in the 1790s when surveyor and adventurer Major Aaron Upjohn tried to pay off his creditor by mortgaging his house, he was unable to produce either a property deed or measurements of the house.[57]

The pattah, therefore, was just one of the many "official" papers attesting ownership circulating in the burgeoning city. Other documents of ownership were wills or entries in the town police records. Often there was simply not much in terms of paper or documentary practices attesting ownership.[58] For instance, alongside the pattah, bills of sale were also

54 Henry Yule and A. C. Burnnell, *Hobson-Jobson, a Glossary of Colloquial Anglo-Indian Words and Phrases, and of Kindred Terms, Etymological, Historical, Geographical and Discursive* (London: J Murray, 1903). http://dsal.uchicago.edu/cgibin/philologic/search3advanced?dbname=hobson&query=*pattah*&matchtype=exact&display=utf8, accessed 27 July 2013.

55 Sterndale, *Calcutta Collectorate*, 36.

56 Saumarez, "Rule-by-Records and Rule-by-Reports," already warns us against reading the records and leases as actual existing conditions of the property market on ground. See also report appended to Extract from the Territories Department, 24 March 1820, Judicial [Criminal], 1 April 1820, Prog. 15, WBSA, Kolkata, which appends a house registry held with the police; for more on renting and tenancy practices and a selection of wills by wealthy Indian families, see also Pradip Sinha, *Calcutta in Urban History* (Calcutta: Firma KLM, 1978), 1–32 and 140–59.

57 Peter Robb, "Mr. Upjohn's Debt: Money and Friendship in Early Colonial Calcutta," *Modern Asian Studies* 47, no. 4 (2013): 1185–271, 1214.

58 In my archival research into property disputes, I have come across cases where the proprietor often could not produce any paper document, other than attestation by an influential person stating the nature of ownership. Pradip Sinha's early work also reveals that some but not all wealthy Bengali landlords of Calcutta had wills. Sinha, *The Urban Experience*, 140–59. In the case of the Strand Bank, land ownership was noted in a tabular form appended to the Report. Mariam Dossal gives a similar account of property documentation in Bombay Dossal, *Theatre of Conflict, City of Hope: Mumbai*, 39.

used to claim ownership, until it was decided in 1780 that bills were valid only as equitable title and not complemented by legal conveyance.[59] The Collectors of the Cutcherry floated multiple proposals through the eighteenth century to fabricate the pattah as the sole recognizable title. Finally, following the sale of the land, the owners were required to register their titles as pattah.[60] In spite of the multiple proposals, a clear demarcation of lands under ownership, lease and rents continued to elude the officials in the Collectorate.

The impossibility of enforcing ground rents resulted in a creative array of proposals by the various overzealous collectors who followed John Surman, but the legality of these proposals was so dubious that even the Company of the late eighteenth century, which was beset by corruption, had to brush them aside.[61] On many occasions the Collectors complained to the Board how much revenue the Company lost since proprietors neither applied for a pattah with their office nor registered their sale of lands. The apparent losses from the ground rent were so great that in 1789 Collector Mr. Seton announced that, if the proprietor sold land without paying the necessary taxes, the government would liquidate the land.[62] The proposal to liquidate land acquired through "force and fraud" was never enforced as the Company, gaining in authority and maturity, considered it *ultra vires* (beyond the legal powers of the Company).[63]

As records from the Court of Requests demonstrate, there were many defaulters of ground rent, and the multiple attempts at imposition only resulted in creating a long string of litigations in the Company courts. The Company established a punitive regulation in which the poor and lower caste defaulters were to be flogged, the wealthier ones were to be fined.[64] Even after developing a differentiated punitive system based on caste and class, the Company found itself perpetually struggling either to verify land titles or to force a regulated payment of ground rents from the growing population of the city. The land market, in short, remained mostly beyond the absolute grasp of the Company.

For an account of property litigation in Madras, see Niels Brimnes, "Beyond Colonial Law: Indigenous Litigation and the Contestation of Property in the Mayor's Court in Late Eighteenth-Century Madras," *Modern Asian Studies* 37, no. 3 (2003): 513–50.

59 Legal conveyance is the notarized or registered document that validates the transfer of title from one person to another without any encumbrance.

60 Sterndale, *Calcutta Collectorate*, 39–40.

61 Sterndale, *Calcutta Collectorate*, 41.

62 Sterndale, *Calcutta Collectorate*, 40.

63 Sterndale, *Calcutta Collectorate*, 41.

64 Sterndale, *Calcutta Collectorate*, 39–40.

Finally, the shifting river and its alluvial deposits in the early nineteenth century gave the Company expanded scope of authority vis-à-vis land in the city. The productive powers of fluvial capriciousness were revealed to the Company agents, and they exploited those with great alacrity. A whole host of regulatory measures and a paper-based regime of property were unleashed on the newly formed land. The implementation of these regulatory measures was intended to be an experiment in urban land alienation and transfer of land into the Company coffers that, if successful, would be implemented throughout the city. One of the immediate effects of the alluvion deposition was to delegitimize the first cadaster survey of Calcutta from 1797. Following the proposal for a new survey in 1819, Mr. Lind, the collector of Calcutta, proposed that boundaries of land had to be inserted in all future pattahs.[65] This also emboldened the Advocate-General to order notices soon after, in 1829, requiring occupiers to pay their ground rent directly to the collector or be tried in the Court of Requests. This is how the trading Company fashioned itself as a public agent and trustee of lands in the city.

The rule of law vis-à-vis ground rent and landholding in Calcutta revealed the limits of Company authority. With all kinds of regulatory mechanisms exhausted, the Company finally turned to the "plenitude of prerogative" within colonial law to settle the disordered state of urban landholdings in Calcutta.[66] The shoals and mudflats appearing along the riverbank in what was becoming a burgeoning real-estate market in the early nineteenth century finally enabled the Company to consolidate its authority in land. It did so by embedding its infrastructural projects of embankment and public works into the legal architectures of ownership and stipulating a strict regime of documentary practices. These moves legitimated particular forms of quasi-state domain in land and thereby enabled a form of market governance from as early as the 1820s onward.[67] Ultimately, the Company emerged as, to quote one official, the "general-zamindars" of Calcutta.[68]

What was the nature of this zamindari system? A word of clarification is necessary to parse out the difference between an agrarian zamindar in

[65] Sterndale, *Calcutta Collectorate*, 37.

[66] According to Nasser Hussain the contradiction that structures colonial law was the possible use of emergency in the name of prerogative to subvert the very rule of law that justified British colonialism. See Hussain, *The Jurisprudence of Emergency*, 2 and 16–22.

[67] Birla, *Stages of Capital*, 8–10.

[68] Extract of the Territories Department, 1 April 1820, Judicial [Criminal], Prog. 15, §48, WBSA, Kolkata.

the established scholarship and the zamindari as it related to the factory town of Calcutta. Early Calcutta, around the time of the British settlement, consisted of a number of weekly markets owned by a handful of people who drew taxes for farming out those markets.[69] Through decades of contestation the Company sought to acquire those markets and build settlements around them. The Company's zamindari system in Calcutta was *lakhiraj*, or rent-free tenure, with the limited rights to sell, grant or lease the *coomar*, or unoccupied lands, and to receive a rent not exceeding three rupees sicca per bigha from occupied lands.[70] If a situation arose where any owner failed to pay his arrears to the government, the Company had no rights to resumption or sale, as in the case of permanently settled agricultural lands. Such defaulted arrears were recoverable only through the sale of movable property.

The Collector, a salaried junior civil servant of the Company, was charged with issuing leases or pattahs under the Company's seal for tenure of houses or grounds.[71] From existing records we know that the Collector drew an annual salary of approximately 2,000 rupees from 1720 to 1756, but from extant accounts it also appears that his role was somewhere between a paid tax collector and a plunderer. As Sterndale pointed out, the "bulk of the receipts from the farms appears to have gone into his pocket. Besides this, he was actively engaged in private trade, from which he derived a very large income."[72] Collecting taxes on urban land and imposing ground rent were not systematized until 1861, three years after the Crown took possession of British India, when an act to tax urban property was passed.[73] Thus, unlike forests and agricultural landholdings, urban land was slow to become part of the systematized balance sheet of revenue generation. The unregulated system of tax collection, compounded with corruption at the Cutcherry, continued for more than a century as Calcutta evolved over the 1800s to become the second capital of the empire and an important node connecting the trading worlds of Europe to South and East Asia.[74]

[69] Sen, *Empire of Free Trade*, 49–51. [70] Sterndale, *Calcutta Collectorate*, 32.
[71] Sterndale, *Calcutta Collectorate*, 12. [72] Sterndale, *Calcutta Collectorate*, 13.
[73] "Act for Imposing Duties on Profits Arising from Property, Profession, Trades, Office, Act XXXII of 1860," 19th-Century House of Commons Sessional Papers, vol. XLIII.201, 7 March 1861.
[74] Kapil Raj, "The Historical Anatomy of a Contact Zone: Calcutta in the 18th Century," *Indian Economic and Social History Review* 48, no. 1 (2011): 55–82.

MANUFACTURING A RIVERBANK

Following the shifts in the course of the river in 1804–5, the Conservancy and Sanitation Department of the Justices of Peace was tasked with the responsibility for technologically separating the land from water and fortifying the marshes into a recognizable geographical formation: the riverbank.[75] They carried this out by depositing the offal of the city to prop up the watery soil, which resulted in a cholera outbreak, causing the immediate discontinuance of the project.[76] Apart from government efforts, many of the property-owners along the riverbank began to brace and raise the land between low and high water tides with stones and *surki* (brick dust), hoping to stake a claim to it. Fortification of parts of the riverbank for easy access to the river was not new to India. Wealthy merchants would construct landing areas with steps known as *ghat*. The colonial state found these piecemeal constructions chaotic and wanted to fortify the entire area as an uninterrupted stretch of promenade and road. For instance, Radhamadhab Banerjee, whose house was along the Neemtolla Ghat (a landing area with steps and a crematorium), had raised the newly acquired land in front of his property with stones, *jhamma* (burnt brick), earth, bricks and surki. These privately funded land fortifications gave impetus to the Territories Department to conduct a survey in 1820 in order to ascertain which part of the ground along the river could be considered the property of the Company.[77] The Lottery Committee would then take over the land for the purposes of improvement. As in the case of Banerjee, the Lottery Committee took over the land in 1826 in the name of public purposes, and Banerjee's petition for reimbursement remained unsuccessful.[78]

Through the Report published in 1820, the Territories Department recommended improvements in the riverbank area that significantly impacted property holding in Calcutta. Initially, the Department began the process of establishing the Strand Bank Fund, a provisional body established to raise money to improve the riverbank area. However, the Strand Bank Fund did not begin functioning until 1837, a full seventeen

[75] David B. Smith, *Report on the Drainage and Conservancy of Calcutta* (Calcutta: Bengal Secretariat Press, 1869).

[76] For more on this, see my "Fictions of Possession: Property, Law and Capital in Colonial Calcutta" (PhD thesis, Emory University, 2014), 94–95.

[77] Extract from the Proceedings of the Territories Department, Judicial [Criminal], 1 April 1820, Prog. 15, WBSA, Kolkata.

[78] Nimtollah Ghat, Judicial [Criminal] 31 August 1826, Prog. 22, WBSA, Kolkata.

years after this initial recommendation in 1820. There was also vested interest in regularizing the tolls and wharfage on boats docked along the banks and those using the area for loading and unloading, which mostly went to private property owners.[79] Through the efforts of the Strand Bank Fund, the Territories Department gradually appropriated the riverfront lands from the wealthy Indians and private British merchants for beautification, trade and warehouse purposes. No residents with property along the riverbank were ready to give up their land to the Company without a fight, the fiercest among them being a Hindu deity, as will be discussed in Chapter 3.[80]

The Report, however, played a critical role in reconceptualizing the nature of the Company's land ownership in Calcutta.[81] How are we to read this document? It would be too narrow to read the Territories Department Report as a mere fiscal answer to the problem of land ownership in Calcutta. At the same time, it was not simply a set of recommendations. Rather, the Report operated more like an imagined cartography of the city that emplotted land, value and people within the emerging urban space by notarizing property and modes of ownership.[82] Thus, the Report and the subsequent paper trail it generated were the bureaucratic arm of conquest grappling with the nature of the Company's status in Bengal and the question of British power in the empire. Particularly, the Report afforded the Company servants with conceptual categories to delineate river, land and depositions. These delineations of the natural river formations helped generate an administrative language of occupation. In this manner, the Company bureaucratized acquisition of riverine accretions outside normative judicial structures. By ushering paper and the Company's insignia as *the* documentary tool to record property in the city, the Report operated like a seal of notarization that fixed land and land ownership. Indeed, following the recommendations in the Report, the Territories Department was able to abrogate the question of the justice or necessity of taking over land, by making way for a particular kind of imperial legal architecture in urban landownership.

The Report conceptualized the new land formations as public lands with the Company as its trustee, eliding the long history of preexisting

79 Extract from the Proceedings of the Territories Department, Judicial [Criminal], 1 April 1820, Prog. 15, §§49–51 WBSA, Kolkata.

80 Strand Bank Funds, Judicial [Judicial], 10 October 1863, Prog. 123–31, WBSA, Kolkata.

81 Extract of the Territories Department, 1 April 1820, Prog. 15, WBSA, Kolkata.

82 I borrow the concept of imagined cartography and notarizing forms of possession from Smail, *Imaginary Cartographies*, 65–109.

communal, affective and Indic religious investments into lands along the holy river. Nuisance, as we know, has repeatedly been deployed by authorities, whenever it suited its purposes to discipline spaces, populations and behavior.[83] In keeping with this process, the Territories Department along with the Collectorate made a concerted effort to remove "nuisances" along the riverbank, which included huts, shops and temples between Calcutta's Chandpal Ghat and Chitpore (an area comprised mainly of native inhabitants) in order that they might not impede free movement and trade.[84] Such so-called nuisances also included all forms of obstructions strewn about such as garbage, old guns, brick rubbish and timbers.[85] The regulation of public annoyances was premised on much the same political modernity attributed to private property in the Bengal Presidency's more settled agrarian locales. Disciplining techniques along the riverbank by British officials, and in some cases even by the new Indian middle classes, provided the architecture for the ordering of public spaces in such water-soaked resource frontiers.[86] In spite of the disciplinary regime set in place, the Territories Department Report records a slippage in the conceptualization of the ghat as public property or – to be more precise – between what is publicly *owned* versus what is communally *used*. This slippage points to an uneasy relation between the ideas of space and property. Space could not be conflated with the idea of property, just as property did not encompass all the relations that inhered in a space, especially one considered holy by caste Hindus who also happened to be landlords wielding significant power in the burgeoning city.[87]

Having begun the process of removing "obstacles" in 1820, Esquire Holt Mackenzie, the architect of the Report and the Secretary to the Territorial Department, set out to ascertain to what extent the Company

[83] Joel Franklin Brenner, "Nuisance Law and the Industrial Revolution," *Journal of Legal Studies* 3, no. 2 (1974): 403–33.

[84] First Report of the Special Committee for Considering the Nuisances Which Exist throughout the Town of Calcutta and Proposing the Best Means of Removing Them, Judicial [Criminal], 25 July 1805, Prog. 22–25, WBSA, Kolkata.

[85] First Report of the Special Committee for Considering the Nuisances, 25 July 1805, Prog. 23, WBSA, Kolkata.

[86] Michael Anderson, "Public Nuisance and Private Purpose: Policed Environments in British India, 1860–1947," SOAS School of Law Legal Studies Research Paper Series, Research Paper No. 5 (2011) (reprinted from SOAS Law Department, Working Paper No. 1 (July 1992)), available at https://ssrn.com/abstract=1884551.

[87] Chapter 4 looks at parallel projects of wet dock constructions and how the space along the river was completely reorganized according to economic principles of maximization of profit.

could appropriate this newly formed ground and claim it as its property. As an initial step toward this appropriation, the department undertook an extensive survey of the riverbank along Calcutta, which included the newly formed shoals and mudflats, with the goal of claiming them as property.[88] Producing a documentary basis and legal argument for two forms of possession within the empire preoccupied Mackenzie from 1819 to 1822. One of these forms of possession was the agrarian revenue settlement he was carrying out on one of the most fertile lands of the Ganga Delta in the northern provinces. Also known as the Mahalwari system of land settlement, this space was an admixture of both the Permanent Settlement in Bengal and Thomas Munro's Ryotwari system in the southern Presidencies.[89]

The second preoccupation was Mackenzie's attempt to fix through law these land–water admixtures in order to secure a quasi-state domain in land in nonagrarian space. As we shall see, this process of transforming these aquatic resources into acquisitionable property was part of the technique of securing the materiality of the paper as a legitimizing object of modern property. As a governing document or a "graphic-artifact," the Territories Department Report mediated and transformed many social relations into political and economic contracts and it became a significant part of the nineteenth-century social life of the city.[90]

The Report stressed that in Calcutta the Company had the *double right* of both the sovereign and the zamindar. Therefore, it harnessed the laws of the sovereign alongside established custom. The availability of the newly formed alluvium land along the river, thus opened up a variety of claims. Mackenzie stressed throughout the Report that "[t]he company is the general Zamindar and as such any new lands must, we conceive, be considered at its disposal."[91] However, such was not the practice by the Company even just some thirty years prior. A fitting example is a case regarding accreted urban land that came to the Court of Request in 1789.

88 Extract from the Proceedings of the Territories Department, Judicial [Criminal], 1 April 1820, Prog. 15, WBSA, Kolkata.

89 Romesh Chunder Dutt, *The Economic History of India under Early British Rule, from the Rise of the British Power in 1757 to the Accession of Queen Victoria in 1837* (London: Tübner, 1906), 190–94.

90 I am borrowing from Mathew Hull's formulation to explain how paper (both as a linguistic text and as a non- and paralinguistic entity) mediates the act of governing with citizens' acquiescence, contestation and use of governance. Hull, *Government of Paper*, see especially the introduction and chapters 1 and 4. See also Raman, *Document Raj*.

91 Extract of the Territories Department, 1 April 1820, Judicial [Criminal], Prog. 15, §48, WBSA, Kolkata.

A *sheristadar* (registry keeper) by the name of Jureeff Ameen reported to the Board that the prevalent customary practice in Calcutta was as follows: "when there were land accretions no one could take possession of the land without the permission of the Government, and when the Government had occasion for it, no other persons can have it. But when this was not the case the pottah holder of the adjoining lands has a preferable claim to anyone else."[92] This was how cases about agrarian chars (shoals) were also settled and would be emblazoned into the 1825 Regulation XI on Alluvion within five years after the publication of the Report under discussion.[93]

However, Mackenzie set in motion a process whereby he could revise these existing practices, creating new provisions and thereby setting precedence for acquiring Calcutta's shoals. This process of land acquisition had a thriving life in the city outside the judicial dictates of the Bengal Codes. His report emphasized "that even in the case of persons who have long had possession it is the practice for them when the excess is discovered to apply for and obtain a pattah from the Collector for such excess and pay corresponding Revenue."[94] However, as the next sentence of the report reveals, the text is shot through with doubts of ownership, or – more precisely – with an attempt to forge, in the dual sense of creating and fabricating, a language of ownership. The Report recommended that the Strand Bank authorities should lay "immediate claims" to the newly acquired land exposed "by the spontaneous desertion of the stream, and [that the Company] acquired lands were to be converted to a public wharf from Customs House to Cuchagoody Ghat."[95] The Territories Department hastened to create what it considered would be a "public wharf," but for that the Department needed to demarcate spaces already in use by the public. If successful, Mackenzie added, they could then begin to acquire other similar lands in the city beyond the purview of this survey.[96]

[92] Sterndale, *Calcutta Collectorate*, 40. This very much matches what happened to agrarian chars.

[93] Sinha, "Fluvial Landscape."

[94] Extract of the Territories Department, 1 April 1820, Judicial [Criminal], Prog. 15, §30, WBSA, Kolkata.

[95] Extract of the Territories Department, 1 April 1820, Judicial [Criminal], Prog. 15, §8, WBSA, Kolkata.

[96] Extract of the Territories Department, 1 April 1820, Judicial [Criminal], Prog. 15, §72, WBSA, Kolkata.

Having planted Fort William in the riverfront area, it was deemed that the ghats around it were chiefly used for military purposes, housing yards and buildings for the reception of garrison stores. According to the Report, Babughat, one of the ghats adjoining the fort and an important point of transport between Calcutta and its adjoining areas, was particularly convenient for the embankment of troops and the loading and unloading of the military garrison. Devoid of any form of circumspection about town planning, the area from the north of the Fort to that of Chandpal Ghat was demarcated to be kept clear, and they planned a public promenade. However, the British soon realized the fallacy of creating a "public" space out of a public space, and they covered up their fatuousness through a plan that forbade "private individuals" to enter or use that area until the promenade was ready.[97] Judiciously, the term "private individuals" was left undefined. Yet it seems that nowhere was there any discussion about the necessity and legitimacy of taking this land as Company land. Elsewhere scholars of agrarian riverine landscapes of the Gangetic plain have variously shown that, for newly accreted lands, the Company calculated a revenue structure based on yield, rather than land measurements, to maximize its gains, but did not necessarily make a claim for the land as Company land. Thus, we can safely conclude, for nonagricultural lands, the Company began exercising an exceptional power in seizing any land newly formed by the shifting river and thus refashioning itself as a "land developer."[98]

This particular early history of the colonial encounter with aquatic resource frontiers in Bengal, thus provides a little-known chapter in a well-known story of the British transfiguration of heterogeneous communal spaces into clearly demarcated public spaces, and in the codification of their inhabitants' customs into a legal system.[99] This codification, moreover, was also partly coterminous with the production of a market in land, where the market masqueraded as the colonial idea of the "public," one that could be regulated, measured and disciplined.[100] As the Report further claims, although the government stood to accrue a considerable amount of revenue through the acquisition of this land, in the context of aquatic "resource frontiers" like the

[97] Extract of the Territories Department, 1 April 1820, Judicial [Criminal], Prog. 15, §5, WBSA, Kolkata.

[98] Raman, "Sovereignty, Property and Colonization."

[99] Bernard S. Cohn, *Colonialism and Its Forms of Knowledge: The British in India* (Princeton, NJ: Princeton University Press, 1996).

[100] Birla, *Stages of Capital*, 10.

Hooghly's riverbank, "the immediate pecuniary gain" was hardly the driving force behind the Strand Bank project.[101]

Impulses much larger than mere pecuniary gains were at work here. The Territories Department Report provides compelling evidence for the frustration encountered in the colonial state's early attempts at mapping in the face of the moving river, which constantly rendered earlier ground measurements obsolete as in the case of the 1797 survey.[102] The major thrust of the Report was to acquire lands that the Territories Department found to be areas not subjected to the regime of paper documentation with the pattah office. If inhabitants could not produce any pattah, that meant that they could not produce any "recognizable" land titles. Thus, these spaces within the city emerged from water already entangled in a new language of pattah ownership that secured for the Company rights to take land. This facilitated land fortification as the Company's powers expanded over the nineteenth century, irrespective of the fact that the ecological dynamism of a delta might submerge these lands or other parts of the floodplains that surrounded Calcutta. Whenever residents, mostly native but sometimes also Europeans, failed to supply the surveyors with paper documentation, the officials declared:

> We are of the opinion that it is indisputably the property of the State and it would be advisable to adopt measures for raising it with a view to the appropriation of it to public purposes. We are not aware that any claim is likely to be seriously maintained to this spot, though it has been occasionally used for the deposit of old guns and anchor, and thus a *dubious sort of occupancy* has been exercised over a part of it by Messer's Clarke and Co. on behalf of Mr. Johnson. It may nevertheless be proper to state the grounds on which we consider the title of Government to it to rest.[103]

At this point in time in the 1820s, paper documentation of property became central in illuminating how the operations of colonial law and economy merged in a moment of regularization. This report crystallized the moment of struggle in translating "squatting" or occupation by early British merchants into ownership, or – more precisely stated – colonial

[101] Extract of the Territories Department, 1 April 1820, Judicial [Criminal], Prog. 15, §65, WBSA.

[102] Extract of the Territories Department, 1 April 1820, Judicial [Criminal], Prog. 15, §§11, 13, 15, 16, 17 and 21, WBSA, Kolkata.

[103] Extract of the Territories Department, 1 April 1820, Judicial [Criminal], Prog. 15, §§27–28, WBSA, Kolkata (emphasis mine).

conquest.[104] Early British writings about life in Calcutta often use the term "squatting" to describe the British presence in Calcutta, and the Report clearly states that this survey must seal the translation from squatting to conquest through the redefinition of the idea of legality when it came to grabbing land.

The Report prescribed an arbitrary provision for applying to have the extra land counted as part of existing ownership. It stipulated that papers legitimizing ownership must be produced to justify claims on the newly emerged land. However, by a sleight of hand, it foreclosed even the seemingly suspect provision in the next sentence by saying that, in spite of any legitimate claim one may produce, the "state" may decide to take over the land or refuse to grant rights to it.[105]

Turning to the specific history of this urban marsh reveals how the colonial state juridically intervened into these land–water admixtures since their status as land, and thereby (landed) property, were ambiguous. The report reinscribed the authority of the pattah – land register – into ownership titles. Thus, what was considered the work of a capricious environment unleashed the emergency provisions within colonial regulations to acquire these newly formed aquatic lands as Company lands. After much conflict, the present Strand Road that runs through the city was constructed between 1852 and 1873 "by using the accreted land permanently, and exclusively, for purposes of public utility connected with the trade, the traffic, the health and the convenience of the community."[106] The process of producing the land along the ghat as public property with stipulated land-use regulation was anything but easy and resulted in multiple disputes and regulations. The following chapter will document the afterlife of the report to recover how existing land-use patterns became expressions of fictional claims to property and were rendered recalcitrant in the domain of law.

The Strand Road venture facilitated by accreted lands had a long gestation period in multiple administrative units moving from Territories

[104] For the collusion between mapping and conquest, see Kapil Raj, "Circulation and the Emergence of Modern Mapping: Great Britain and Early Colonial India, 1764–1820," in *Society and Circulation: Mobile People and Itinerant Cultures 1750–1950*, ed. Claude Markowitz (Delhi: Permanent Black, 2003), 23–54; Benton, *A Search for Sovereignty*; Raymond Craib, *Cartographic Mexico: A History of State Fixation and Fugitive Landscapes* (Durham, NC: Duke University Press, 2004).

[105] Extract from the Proceedings of the Territories Department, 1 April 1820, Prog. 15, §31, WBSA, Kolkata.

[106] Chakraborty and De, *Select Documents on Calcutta*, 95.

and Sanitary departments to judicial committees and courts. One important water regulation had a significant role to play in this. Four years after the publication of the 1820 Report, when the Governor-General in Council passed Bengal Regulation I regarding land acquisition, he simultaneously passed another regulation dealing with the aquatic resource frontiers in the active delta: *Regulation for declaring the rules to be observed in determining claims to Land gained by Alluvion or by Dereliction of a River or the Sea*. Also known as Regulation XI of 1825, this regulation laid down strict stipulations based on "local customs" with regard to land gained through alluvion. It stated that the new land belonged to the proprietors of estates in the contiguous areas, and nowhere did this law leave any provisions for state resumptions of newly formed lands. The only stipulation for state domain appeared in the case of char, or islands thrown up by large navigable riverbeds. When such newly formed islands were not fordable, the regulation decreed that it would become government land.[107]

The new strip of land deposited by the river along Calcutta's western banks was not an island, was contiguous with the estates of the people living along the river and should have been arbitrated under the customs of dhardhura governing fluvial borders. Regulation XI building on the dhardhura custom stipulated that "the Court of Justice in deciding upon such claims and disputes [arising from land gained by alluvion or by dereliction of a river or sea] shall be guided by the best evidence they may be able to obtain of established local usage, if there be any applicable to the case or, if not, by general principles of equity and justice."[108] Among the various arguments put forward by the Court of Justice in Calcutta in the drafting of Regulation XI 1825 on Alluvion, one geological clause is salient for our discussion. This clause deemed that customs were place-based and developed in response to the specific geological nature of the land and waterscape. Therefore, examples of riverine behavior from upper Bengal could not be used as a valid legal argument to arbitrate on a case taking place in lower Bengal.[109] This clause gave the Company the right to ignore earlier precedence in the name of geographical specificity. Ecological variability

[107] Field, *Bengal Code*, 586. [108] Field, *Bengal Code*, 592.

[109] In one case an appellant tried to make an argument for a custom prevalent on the banks of Gogra River to be used to adjudicate a case on the banks of the Ganga River, and the case was rejected. *Rai Manik Chand v. Madhoram and Others*, XIII Moo. In Ap. 1 and III of Bengal Law Reports, Privy Council 5.

of the Bengal Delta nullified the reliance on legal precedence in litigating disputes over fluvial borders. Thus, differences between the lower and the upper delta became moot in handling the cases pertaining to the production of the Strand Bank in Calcutta. Embedding the clause as part of the Regulation meant that the Company could keep its adherence to existing "customs," while at the same time dispensing with any need to respect them.

Local evidence or not, in the case of nonusufruct land, the Company's regulation created an opening for taking land through emergency provisions legitimized through the payment or, rather, promise of compensation.[110] Thus, the 1820 Report noted that the "situation [of the new land] and its extent renders it peculiarly valuable, ... though it may perhaps be recovered without necessarily involving the general questions of the government right to alluvion land, or those of the public to the space lying between high and low water mark."[111] The Report further stated that, should such questions arise, the provisions for documentation and "the conditions of the Grant" laid out in the Report will facilitate the decision to take over this strip of land. These contrary provisions in the Report remained valid in spite of the Regulation XI 1825. With this ambivalent coming of the documentary regime in the wake of the newly deposited alluvion, Indian proprietors found themselves tenants in their own land.[112] As we will see in the next chapter, even when it was promised to the proprietors, compensation never came.

Compensation might have secured a dubious legitimacy for the act of taking land. Yet while compensation lent a so-called procedural transparency, it did not diminish the exercise of exceptional powers by the state, which managed to dispense with both an argument of "necessity" for taking land, and any respect for existing practices around arbitrating the emergence of new lands. The politics of taking land played out within the colonial urban context by eliminating the language of

110 Elsewhere I have explored how payments of compensation legitimated land-grabbing under colonialism. See my "The History of Eminent Domain in British Colonial Thought and Legal Practice in South Asia," *Economic and Political Weekly* 50, no. 50 (Dec. 2015): 45–53.

111 Extract from the Proceedings of the Territories Department, 1 April 1820, Prog. 15, §68, WBSA, Kolkata.

112 The ambivalent place of the pattah in the post Permanent Settlement agrarian world has been recorded by Wilson. See Wilson, "Governing Power of Proprietors," in *Domination of Strangers*, 104–32.

necessity.[113] The land accretions and depositions by a tidal river became the grounds for enabling the Company to acquire more land, without taking recourse to any argument for public purposes or "just" compensation. The Company continued to devise laws that would expand its power over the ownership practices in the city. Many of the major urban infrastructural projects were thereafter made possible through emergency provisions to acquire so-called indeterminate landscapes, which are neither bounded land nor flowing water.

MUDDYING PROPERTY

As my analysis of the 1820 Report reveals, the absence of the pattah did not easily translate into a nonoccupation. Indeed, the Report had to reinvent the authority of the pattah in order to demarcate state authority over these aquatic frontiers. Soil, mud and soaking landscape were transformed by the Report and its fabrication of the pattah into property that could be exchanged in a state-regulated market. A malleable juridical reasoning eased the process of land-grabbing by the Company under the guise of extensive infrastructural projects. Force was turned into legal authority and shaped the Company's relation to the urban landscape. Studying the manufacturing of a riverbank in Calcutta reveals the economic ideologies driving the expansion of public works, and the legal logic of securing the newly formed spaces as Company lands from the early decades of the nineteenth century. The land acquisition battles slowly created a thriving market in urban property in colonial Calcutta. Regulating this market served economic purposes of revenue generation and administrative purposes of governance and spatial control of the native population.

This chapter documented the dual process of environmental consolidation through public works and legal maneuvers as the East India Company's practices of capital accumulation were transformed over the course of the nineteenth century. The Company embedded the ecological specificity of the estuarine behavior into the process of property-making in riverine shoals in the aid of extracting value from what became the Company's newly acquired aqueous possessions. Property in urban land, marshes and water began to emerge not just as a legal idea, but also as a

[113] Daniel Skinner and Leonard Feldman, "Eminent Domain and the Rhetorical Construction of Sovereign Necessity," *Law, Culture, and the Humanities* 10, no. 3 (2015): 393–413.

technology to harness the landscape, and as a bureaucratic language to standardize property's movement through the market. Thus, property became a particular way of seeing *land* and entitlement, where there might have been no land in the first place.[114] As the following chapter will demonstrate, numerous petitions followed the construction of a public road along the Strand Bank on accreted land, and many landowners resisted the Company's bid to grab their property.[115] The Company won most of these cases. The everyday lived histories of the contestations and surveys documented in the following chapter shows how the mobility of the landscape was harnessed within the domain of law.

[114] Carol Rose, "Seeing Property," in *Property and Persuasion: Essays on the History, Theory and Rhetoric of Ownership* (Boulder, CO: Westview Press, 1994), 267–304.

[115] Mallicks versus Company Government, Judicial [Criminal], 4 September 1818, Prog. 6–10, WBSA, Kolkata; Petition to Save the Sanctity of Religious Spaces, Judicial [Criminal], 22 August 1822, Prog. 35–41, WBSA, Kolkata.

PART II

LEGAL MANEUVERS

3

Notarizing Possessions

Ecological difference entered the domain of law in colonial Bengal as a legal exception. This chapter documents the mutations that occurred within colonial law after the haphazard application of legal doctrines of acquisition to newly formed lands. As the city officials decided to turn the communally used space of the riverbank into a "public" road, multiple forms of resistance broke out within the city, finally ending at the court of law. The dispensation of justice, especially through British courts and the rule of law, had by the second half of the nineteenth century become the basis for the moral legitimacy of British presence in India. Yet, as we saw in the previous chapter, the institutional practice of law fell short of its own standards, especially since it widened the gap between the necessary and possible juridical arguments. Producing Company lands in the newly formed riverbank brought forth a slew of cases as many of the landowners disputed and resisted the Company's authority to take land. Among those who demanded a just settlement to their proprietary claims was a Hindu deity who landed up in the British courts. As a result of such cases, implementing the proposals in the Strand Bank Report could not proceed unless the proprietary claims were settled.

From 1864 to 1870, a subcult Hindu deity enshrined on the riverbank was a plaintiff at the Privy Council in colonial Calcutta, waging and winning a protracted judicial battle over a plot of land along the newly formed riverbank.[1] This was the only Strand Bank case that the colonial powers lost. The deity and its claim to the riverine deposition, expressed

[1] Legal Tussle with Strand Bank, Judicial [Judicial], 1 June 1870. Prog. 161, West Bengal State Archive, Kolkata (hereafter WBSA).

through the sudden and "surreptitious" appearance of a stone on the riverbank, reveal the irreducible difference that made up the world of resistance produced from the entanglement of nature, law and colonial infrastructure. This case was not about establishing truth claims, nor about countering imagined events with real evidence.[2] Rather, it was about setting ethical limits on the Company's expanding land hunger. The consolidation of an urban economic geography along the muddy waters of the riverbank produced unique forms of ethico-political claim-making on the part of the urban population, which ranged from colonial entrepreneurs, Indian landlords and Hindu subcult deities.

While this peculiar encounter between the colonial state in British India and a Hindu god terminated in the arena of the judicial court, it began along the Hooghly River and is deeply enmeshed in the history of the lower Ganga Delta region of the Bay of Bengal, where law, riverine moods, ancestral deities and their possessions became part of this fluvial narrative of urbanization. This chapter takes up the challenge of reading the interaction between the bureaucratic world of colonial law and the mythic figuration of the deity, neither by explaining away the fantastic, nor by normalizing the enchantment. Rather, the chapter engages the multiple contexts this encounter brings forth for the historian.[3]

It would be easy to read the seemingly impossible encounter between a divine figure and the mundane world of the British courts as just another instance of strategic articulation of resistance by the subject population, where the deity becomes a political symbol of resistance. However, that would elide the rich history of the space where land and water meet. While the domain of the sacral continues to be a site of resistance against infrastructural programs of the state, such resistance based on religious symbolism is not a timeless and ahistorical act continuing unchanged through the centuries.[4] Moreover, this does not simply represent a

[2] Gyan Prakash's reading of the incommensurablity between the precolonial world of *kamia-malik* (laborers–landlord relationship) as irreducibly different from the world of capitalist landlord and laborers is a useful reminder here to remain alert to those moments in the historical archive where the relations between "real" and "imaginary" events cannot be reduced to one of truth claims. *Bonded Labor*, 34–81.

[3] I reread the various origin narratives of the name of Calcutta in order to explore the ways of disrupting the boundaries between factual, remembered and mythic pasts. Bhattacharyya, "Geography's Myth."

[4] Anthropologists have explored how subaltern populations, in particular, use religious symbolism to protest urban infrastructural projects. In a more recent study, Nicolas Jaoul shows how Ambedkar or Buddha's images, statues and photos are mobilized by Dalit slum-dwellers in order to thwart eviction. Nicolas Jaoul, "Learning the Use of Symbolic

strategic instance of insurgency. Indeed, an almanac attunement to the riverbank reveals the multiple meanings that make up such spaces of dense habitations and helps us recover the sites of overlap between the mercantile and the cosmological orders. Parsing out the many motivations that gathered in these riverine depositions and understanding the colonial logic of organizing them into units of productivity reveals how the British negotiated these many overlapping layers. This chapter analyzes the property dispute cases and their divergent resolutions to highlight how meanings of riverine land were transformed over a century. Simultaneously, it documents the malleable spectrum of the colonial state's justificatory logic for taking land in the name of urban infrastructure. While the preceding chapters treat the river as a site of confluence with land, this chapter deals with the river in both its material form and the metaphoric meanings it had for the people living along and with it.

WRITING HISTORY WITH THE RIVER

The European geographical lexicon, which contains a rich repertoire of tropes in travelogue writing, cartography and landscape painting, depicted riverine regions in the colony as both a natural setting and the space of difference in which the initial moment of the colonial encounter with the indigenous world unfolded. Throughout the nineteenth century, managing the aqueous spaces was at the heart of an emerging "territorial consciousness," which simultaneously reduced nature to a single "natural category."[5] This reduction was legal, technical and aesthetic. The way in which rivers and their banks feature in colonial writings and nineteenth-century colonial paintings created a specific lexical index for viewing colonial riverbanks.[6] For instance, one of the more iconic and haunting literary images of such a colonial encounter was Joseph Conrad's narration of Marlow's upstream journey along the River Congo as a descent into madness. Marlow's inability

Means: Dalits, Ambedkar Statues and the State in Uttar Pradesh," *Contributions to Indian Sociology* 40, no. 2 (2006): 175–207. Arjun Appadurai, "Deep Democracy: Urban Governmentality and the Horizon of Politics," *Public Culture* 14, no. 1 (2002): 21–47.

[5] Alice Ingold argues that managing water was central to the emergence of a particular territorial consciousness in nineteenth-century Europe. See Ingold, "To Historicize or Naturalize Nature: Hydraulic Communities and Administrative States in Nineteenth-Century Europe," *French Historical Studies* 32, no. 3 (2009): 129–53.

[6] On the role of colonial pictorial repertoire consisting of paintings, lithographs and photographs, see Christopher Pinney, *Camera Indica: The Social Life of Indian Photographs* (Chicago, IL: University of Chicago Press, 1997).

to relate to the colonial topography in *The Heart of Darkness* is echoed in another geographical setting: Adela Quested's disorienting encounter in the Malabar caves in E. M. Forster's *The Passage to India.*[7]

Coastlines and riverbanks have been the sites of colonial encounter in many parts of the world, producing a rich pictorial archive.[8] European travelers viewed the "marvelous" world of the colonies as coastlines of wonder and difference, as Charles Joseph's "Topographical Survey of the River Hooghly" demonstrates:

> And it was at this Chandpal Ghat that the first Judges of the Supreme Court, who came out to redress the wrongs of India, but created infinitely more mischief than they remedied, first set foot in India. It was here, at this *Ghat*, that the Chief Justice, as he contemplated the bare legs and feet of the multitude who crowded to witness their advent, exclaimed to his colleague, see Brother, the wretched victims of tyranny. The Crown Court was not surely established before it was needed. I trust it will not have been in operation six months before we shall see these poor creatures comfortably clothed in shoes and stockings.[9]

Legal mechanisms, Joseph hoped, would establish order in the muddy knee-deep swamps of the Hooghly riverbank. This bureaucratic document partakes in the administrative language of the survey, while borrowing a particular trope from colonial travel writing, which describes the sighting of the land after months of weary travel as an encounter with the "other" of the colony. As Lauren Benton points out: "Voyage chronicles typically offer only brief descriptions of the sea, and land sightings and interactions with locals figure as the dramatic turning points of these narratives, and as contrast to the monotony of ocean travel."[10] Such travel writing had more than simply a literary function and also served as reports back to the metropole. European travel writing about India abounds with similar narratives of sightings of the coastline as a moment of wonder and reportage.[11] Paul Carter captures the epistemic

[7] Joseph Conrad, *Heart of Darkness* (New York: Penguin, 2007 [1899]); E. M. Forster, *A Passage to India* (London: Edward Arnold and Co., 1924).

[8] Here and throughout the chapter my ability to read the mimetic capital of representation and the nature of representational practices comes out of an engagement with Stephen Greenblatt, *Marvelous Possessions: The Wonder of the New World* (Chicago, IL: University of Chicago Press, 1991).

[9] Charles Joseph, "Topographical Survey of the River Hooghly from Bandel to Garden Reach, Exhibiting the Principal Buildings, Ghats and Temples on Both Banks, Executed in the Year 1841," *Calcutta Review* 6, no. 3 (January–June 1845): 428–62, 433.

[10] Benton, *A Search for Sovereignty*, 105.

[11] Maria Graham, *Journal of Residence in India* (Edinburgh: Archibald Constable and Co., 1812); Mary Sherwood, *Life of Mrs. Sherwood with Extracts from Mr. Sherwood's*

consequences of this moment of encounter for geographical science and history writing, describing "the coastline [as] an artifact of linear thinking, a binary abstraction that corresponds to nothing in nature."[12]

The previous two chapters explained the bureaucratic and legal battle of producing this abstract hard line dividing the river from the city and its infrastructure, even where no such division existed. Rivers, in the hands of the Company merchants and colonial officials, became historically contingent entities that were expected to flow with a measurable regularity.[13] Throughout the nineteenth century the colonial hydrographers attempted to produce knowledge about water by measuring rainfall, charting tidal patterns, mapping shifts in river's movements and approximating depth.[14] Estimating the behavior of rivers was central to controlling the distribution of water, while also predicting and controlling the river's breaching of the hard cartographic lines of its riverbanks.[15]

In important trading cities across the world, rivers have functioned variously as natural highways and sewers, but also as sources of power and sites of aspiration. Existing within a variegated urban history of connections and uses, these rivers etch trails of stories of loss and dreams.[16] Riverfronts are the point through which the river can be read

Journal during His Imprisonment in France and Residence in India (London: Darton 1857), 246–47.

12 Carter, *Dark Writing*, 9.

13 Just as the idea of fixity in land is not a given, the idea of the river as a flow was also constructed as rivers across Europe were canalized. Riverine water could not and should not stand still. As a means of transport, river water had to flow, not seep the landscape around. Cosgrove and Petts, *Water, Engineering and Landscape*, 1–11.

14 See for instance Barlow's copious notes on tides, Private Papers of Barlow IOR/X/9128.

15 Producing scientific truth about water through hydrology, metereology and oceanography was on the rise across the world from the late eighteenth century. See for instance Helen M. Roswadowski, *Fathoming the Ocean: The Discovery and Exploration of the Deep Sea* (Cambridge, MA: Belknap, 2005); Michael S. Reidy, *Tides of History: Ocean Science and Her Majesty's Navy* (Chicago, IL: University of Chicago Press, 2008); Katherine Anderson, *Predicting the Weather: Victorians and the Science of Meteorology* (Chicago, IL: University of Chicago Press, 2005).

16 For works that deal with the riverine landscape of Bengal, see for instance Himanshu Prabha Ray, "The Archaeology of Bengal: Trading Networks, Cultural Identities," *Journal of Economic and Social History of the Orient* 49, no. 1 (2006): 68–95; for works that deal with rivers as a cultural form in histories of major cities of world, see Peter Ackroyd, *Thames: The Sacred River* (London: Chatto & Windus, 2007); Morris, *Big Muddy*; Kelman, *The River and Its City*; Douglas H. Johnson, "Reconstructing a History of Local Floods in the Upper Nile Region of the Sudan," *International Journal of African Historical Studies* 25, no. 3 (1992): 607–49; el-Sayed el-Aswad, "Spiritual Genealogy: Sufism and Saintly Places in the Nile Delta," *International Journal of Middle Eastern Studies* 38, no. 4 (2006): 501–18.

and written into the city as both a "natural artifact and [a] cultural form."[17] Writing about the British arrival on the banks of the Hooghly River in his *Early Annals* (1895), C. R. Wilson waxes poetic about Hooghly's turbulence:

It [an Indian river] approaches its end rich with spoils gathered during a long and prosperous life, but uncertain where to leave them. Torn in a hundred different directions, it reaches the sea through ever-varying distributaries. Now the *stream eats away its right bank, now its left.* It oscillates in wide sweeping circles, depositing silt on either side, and again breaks through the curves thus formed and takes a more direct course.[18]

While we have a rich and significant body of work on rivers, in most of these writings rivers are read as an unbroken line on a map with measurable movements breaching the banks.[19] There was, and continues to be, faith in the measurement and approximation of these movements.[20] Thus, when the hungry tides of the Hooghly fed upon the banks and churned up new lands, colonial officials understood those movements as a breach that obfuscated attempts by colonial topographers to lexicalize the tidal vagaries.

While lawyers and bureaucrats struggled with the linguistic frames to categorize riverine behavior, some early cartographers, including James Rennell, were nonetheless aware of the continuous process of land formation and subsidence in the Bay of Bengal. This awareness is demonstrated in the works of Italian cartographer Giacomo Gastaldi, *India Tercera Nova Tabula* (1548); the Portuguese historian João de Barros, *Décadas da Ásia* (1552); navigational maps by the Dutch merchant Pieter van den Broecke from 1666 and James Rennell's *A Map of Bengal, Bahar, Oude and Allahabad* (1786).[21] By viewing their maps and consulting Rennell's survey diaries, we see how these cartographers struggled to render the temporality of the delta into the static frames of their

17 Kelman, *A River and Its City*, 7.

18 Wilson, *Early Annals of the English in Bengal*, 103 (emphasis mine).

19 One exception is Dilip da Cunha's forthcoming book *Invention of Rivers*, which provocatively argues that rivers are a product of a particular visual literacy and shows how much of the Bengal Delta is a rain-fed landscape. See also Mathew Morse Booker, who reads San Francisco's relation with its coast through food, especially oysters, to reveal novel ways of reading how the space of the coast mediates human relations to the waters around them. Booker, "An Eddile Bay," *Down by the Bay*, 111–49.

20 On the history of measurement and modeling, see Paul N. Edwards, *A Vast Machine: Computer Models, Climate Data and the Politics of Global Warming* (Cambridge, MA: MIT Press, 2010), 111–38.

21 Iqbal, *The Bengal Delta*, 3.

cartographic reference. Yet, it would be wrong to say that the aforementioned cartographers were attuned to the temporality of the river. Almanac was not an available frame of reference, for instead of seeing the limits to their representational frames, some like Rennell actively worked in fixing the landscape.

For the majority of the Europeans living and working along the coast, their symbolic repertoire drew on European rivers and was enriched by their own religious, biblical and literary writing, all of which provided no precedence for framing this space. As Wilson further narrates:

> A tidal river, the Hughli has not during the last three or four hundred years so much changed its course, but the alterations which have taken place in its confluence and in its banks have been so many and so considerable that an enquirer into its topography in the days of Job Charnock [considered the founder of Calcutta] will often find the greatest difficulty in tracing out the many localities which were at that time well known and conspicuous.[22]

The natural movement of the temperamental Indian rivers, which brimmed with a "swarm of possibilities" as a site of geographical variety and wonder, simply became an anomaly they sought to fix.[23] Thus, once riverine activities entered the legal and bureaucratic registers, the many tidal behaviors were flattened and made more manageable as a water resource with a predictable flow.[24] Consequently, the shifting banks of the river proved intractable to revenue calculation and to the legal jurisdiction of spaces.

Throughout the nineteenth century the legal language of property and the bureaucratic language of survey struggled to posit human–land–water relations as separate and bounded. However, much remained excluded by those frames of reference.[25] Writing in 1856, a team of British engineers explained to the Governor of Bengal in British India the impossibility of "supposing a margin" in the case of rivers in Bengal. Indeed, the seasonal nature of rivers had made these engineers wonder if some of the channels might have simply been *nala* (drains, spill channels or freshes) rather than

[22] Wilson, *Early Annals of the English in Bengal*, 103. [23] Carter, *Dark Writing*, 1.

[24] Reidy documents the scientific and military impetus behind the early calculation of tides in the river, *Tides of History*, 122–56.

[25] In the Canadian context, Paul Nadasdy writes about a similar struggle in land–animal–human relations that could not be easily translated into a language of property and hunting rights, "'Property' and Aboriginal Land Claims in the Canadian Subarctic: Some Theoretical Considerations," *American Anthropologist* 104, no. 1 (2002): 247–61. In the context of aboriginal Australians, Povinelli has shown how sweat, land, ancestral spirits and people constitute one another mutually, *The Cunning of Recognition*, see esp. chaps. 5–6.

nadi (rivers).[26] This realization occurred only in the middle decades of the nineteenth century after colonial engineers had significantly altered much of the riverine landscape in the Indo-Gangetic plain.

Colonial geologist James Ferguson began mapping the movements of the Gangetic Delta from the mid-nineteenth century. While his findings revealed to him the terminological limits he was working with, these limits failed to register with him. His work remained focused on tracing the changes in riverine patterns that had already occurred and on using the data to develop a model for forecasting future changes in fluvial behavior.[27] Moreover, any doubts about the geographical lexicon, even when expressed, remained somewhat marginal, sidestepped by the forward march of the colonial law, infrastructure and property regime.[28]

By turning to Bengali accounts of the rivers, we can recover vastly different manners of relating to water and land. Bengal's regional identity weaves the river-centric landscape into its understanding of land, water, trees and human life.[29] The char lands that are thrown up by the river one day and eaten away the next are familiar lived environments in the Ganga-Brahmaputra-Meghna region.[30] Muddy lands hold together animals, fish, vegetation, humans and gods in inseparable and interdependent ways. Riverbanks have also been important sites for Hindu temples and Sufi shrines, while remaining embedded in economic networks due to their proximity to trade and pilgrim circuits. Riverbanks in the Gangetic plain are characterized by a landing area with concrete steps leading to the river, forming a communal space for bathing, taking ablutions and a regular market. Hindu cremation sites also dot this landscape.[31]

As confirmed by early modern pilgrimage narratives, the landing areas linking the river to the dried land are pilgrim–market complexes where the

[26] Report on the Embankment of the Rivers of Bengal, 1846, IOR/V/27/732/36, 1–2.

[27] Ferguson, "On Recent Changes in the Delta of the Ganges," 321 and 350–54.

[28] Almost half a century later in 1900, J. G. R. Forlong, a former Superintendent Engineer of Bengal, would predict the disappearance of the coastline in a letter to the editors of the *Royal Geographical Society Journal.* Although part fantasy and part science, many of his predictions ring true in our contemporary moment. However, his forecasts seem to have fallen on deaf ears in the early days of the twentieth century. J. G. R. Forlong, "The Life of Calcutta as a Seaport and the Mercantile Capital of Asia," *Geographical Journal* 16, no. 2 (1900): 238–41.

[29] Radhakamal Mukherjee, *The Changing Face of Bengal: A Study in Riverine Economy* (Calcutta: University of Calcutta, 1938).

[30] For a contemporary literary representation of this space, see Ghosh, *The Hungry Tide.*

[31] Diana L. Eck, *Banaras: City of Light* (London: Routledge & Kegan Paul, 1983).

religious and secular intertwine.[32] The logic of the market is also deeply embedded in the sacred rituals, rites and exchange of gifts that take place along the riverbanks. In these narratives, such spaces also acted as arbiters of political power and gateways to important towns. The literary material I analyze in the remainder of the chapter illuminates the political economy of circulation of this volatile geography, rather than a codified and fixed possession backed by the bureaucratic machinery of a modern state.

The riverine landscape of Bengal is woven into its cultural sphere through folk songs, ritual performances, literature, artistic and artisanal representations. Hydraulic and riverine metaphors are central to the cultural and philosophical landscape of Bengali literary works, especially as depicted in the various genres of songs and anecdotes in Bengal.[33] In these myriad representations, journeys along the river often become a metaphor for life. In order to understand the cultural meanings of the river and its banks, it is perhaps best to turn to the unique genre of precolonial didactic narrative poems about river journeys called the *Mangalkavya*. These literary narratives teach the advantages of worshipping particular deities and the ameliorative benefits (*mangal*) that the worshippers might achieve through these practices. These poems were in oral circulation in various parts of rural Bengal from the fifteenth century onward, during the period of rapid expansion of the agrarian frontiers under the Mughal rule. They were recorded beginning in the latter half of the eighteenth century.[34] While historians of Bengal have yet to mine these rich textual sources, literary scholars frequently argue that the *Mangalkavya* poems attest to the birth of a mercantile culture in Bengal from the fifteenth century on.[35] David Curley shows that these texts are

32 See Sen's *Tirthamangal* for an account of the early modern life of the ports along the river, which were both sacred, mercantile and yet eagerly opening up to the changes brought by European trade. A contrasting traditional port history is found in Nilmani Mukherjee, *The Port of Calcutta: A Short History* (Calcutta: Commissioners of Port of Calcutta, 1968).

33 For the importance of river in Bengali folk music, see Carola Erika Lorea, "*Learning to Swim in the River of Desire*: Bhaba Pagla's Songs in Their Performative Context" (PhD thesis, University of Rome, 2014); Glen A. Hayes, "Boating upon the Crooked River: Cosmophysiological Soteriologies in the Vaisnava Sahajiya Tradition of Medieval Bengal," in *Shaping Bengali Worlds, Public and Private*, ed. Tony K. Stewart (East Lansing, MI: Asian Studies Center, Michigan State University, 1989), 29–35.

34 Eaton, *The Rise of Islam*, 213.

35 For a discussion of the historian's approach to the literary world of mangalkavyas, see David Curley, "A Historian's Introduction to Reading Mangal-Kabya," in *Poetry and History: Bengali Mangal-Kabya and Social Change in Precolonial Bengal* (Delhi: Chronicle Books, 2008), 1–36. For an exception to Curley's assessment of the historiographic

repositories of political thought from below and they often depict imperfectly governed spaces, while also showing how imperfections can be ameliorated by following the instructions in the kavya texts.[36] As a precolonial archival source, the poems of the *Mangalkavya* genre attest to the complexity of the space of the ghat where the river and its banks are depicted as sites of both worship and trade.

Mangalkavyas ' central theme eulogizes Hindu subcult deities such as Sitala, the goddess of smallpox (popular mainly in the nineteenth century); Chandi, the slayer of demons; and Manasa, the goddess of snakes. Market scenes along the riverbank are an important trope within the kavya texts, albeit frequently of a nature that challenges our realistic-representative index. They hint at a folk conceptualization of the space that accommodates the reticular logic of tides, where markets are firmly located in a cosmological spatial order. For example, a very common narrative thread running through the *Chandimangal* texts is a scene that opens with a Hindu burning ghat serving as a marketplace. It is a market of dismembered human bodies (*pret-bajar*), where the entourage of the goddess Chandi engages in a trade in human body parts.[37] In the folk songs, *bajar*, or market, also serves as a metaphor for the material and earthly aspects of life. The material and spiritual are inseparable within this literary world. The market that is invoked is not one of contracts, but instead one of exchanges governed by what Sudipta Sen terms "proprietary gestures," gestures that incorporate realms of need, desire and material fulfillment.[38]

While the *Mangalkavya* depicted the sacred/mercantile world of the riverbanks that challenged the neat spatial jurisdictions between land and river, Bengali folk songs, which are primarily riverine, evoke a distinct metaphorical understanding of these land–water spaces. Given Bengal's riverine culture, symbolically rivers were never considered separate from land, but often treated as a continuation of dense spaces of habitation of earthly life and beyond. Whenever water appears as distinct from the land mass it marks a moment of disjuncture – an end of earthly life. River crossings in Bengali folk songs imagine habitations in the life beyond the earthly one. It is no surprise that the boat is often a literary metaphor

field, see Kumkum Chatterjee, "Goddess Encounters: Mughals, Monsters and the Goddess in Bengal," *Modern Asian Studies* 47, no. 5 (2013): 1435–87.

36 David Curley, "The 'World of the Text' and Political Thought in Bengali Mangal-Kavya, c. 1500–1700," *Medieval History Journal* 14, no. 2 (2011): 183–211.

37 Sukumar Sen, ed., *Chandimangal* (Delhi: Sahitya Akademi, 1975), 265–68. The text is ascribed to Mukundaram and dated 1589.

38 Sen, *Empire of Free Trade*, 16.

for life and that the boatman is a divine being leading the traveler through the river of life. There are four major groups of riverine folk songs: *bhatiyali-bhawaiya* (tidal songs), *majhigaan* or *sarigaan* (boatmen songs) and *baulgaan* (minstrel songs). Bauls use the river in their ballads as a metaphor for life and death.[39] Witness, for example, this metaphorically rich baul song about crossing the river, redolent with musings on death and crossing to the other life:[40]

> I sit and weep on Bhaba River bank
> I do not know your name boatman
> Whom should I call?
> Who will help me cross?[41]
>
> *Kaandiya akul hoilam*
> *Bhabanadir pare*
> *Majhi tor naam janina*
> *Ami daak deemu kare?*
> *Mon tore ke ba paar kore?*

The body, the space the body occupies, and the immaterialization of that space are all integral parts of this folk-theological and phenomenological understanding of life, the river and the space between the land and river.

The metaphor of the journey characterizes such understandings of the riverfront. These include stories of trade and pilgrimage found in the early-modern texts of the *Mangalkavya* and in the folk songs of the boatmen for whom the river and the ghat are spaces of dwelling, livelihood and the passing away of earthly life. On the contrary, stories of colonial encounter and river journeys are rich instances of knowledge production. They are laden with tropes that cultivate a particular visuality, one characterized by a cartographic-mindedness that captures and

[39] The four major folk songs are *Bhatiyali*, which comes from the term *bhati* or low-lying area that is mostly inundated during the monsoon flood. The rhythm and meter of the songs are said to have derived from the sounds of the river. *Majhigaan* refers to boatmen and are often sung by women when their husbands are away on the river. Boatmen sing *Sarigaan* during boat races. Sukumar Ray, *Folk-Music of Eastern India with Special Reference to Bengal* (Kolkata: Naya Prokash, 1998).

[40] For a discussion on the rich philosophy behind *baul* thought and *deha-tattva* theology, see Sanat Kumar Bose, "Baul Songs of Bengal," in *Folkmusic and Folklore: An Anthology*, vol. 1, ed. Hemango Biswas et. al. (Calcutta: Folkmusic and Folklore Research Institute, 1967), 45–56; Edward Dimock, *The Place of the Hidden Moon: Erotic Mysticism in the Vaisnava-Sahajiyā Cult of Bengal* (Chicago, IL: University of Chicago Press, 1966).

[41] This *baul* song from rural Bengal was popularized in the late twentieth century by Ranen Chaudhury (translation mine).

fixes the spatial mobility. The colonial journey is one of reterritorialization that involves mapping, measuring and fixing.[42] The folk songs, poetry and paintings, on the other hand, recount a journey away from earthly life, thereby rendering the land immaterial and land–water relations deterritorialized. The manner in which the river–land confluence emerges in these folk songs and poetry also alerts us to the ethical limits of fixing nature's mobility through a regime of ownership.

The representational spaces in the kavya texts and folkloric songs hint at the cosmological space that is woven into the geographical locale. These texts encourage us to read them not for their realistic depictions, but instead through the lens of an almanac. The embodied logic and affect that saturates these spaces of habitation are brought to life through an almanac attunement. Thus, in order to understand how the Hindu deity ended up in the colonial court, we need to understand the multidimensional nature of the space or spaces as "perceived, conceived and lived."[43] While such nonrealistic illustrations challenge the secular craft of the historian, it would be a mistake to ignore those aspects that exceed the bounds of our representational genre. Read symbolically, the market scene of dismembered bodies from *Chandimangal* captures the overlapping worlds of exchange, both sacred and mercantile, that the space signified. Artisans of Bengal continue to paint these richly imagined spaces as narrative scrolls that are displayed as an accompaniment to the oral performance of the *Mangalkavya* poems.

Paintings are another rich source illustrating how rivers inhabit different registers in the vernacular and European pictorial imaginaries. While the archives of precolonial paintings of rivers of Bengal are hard to locate, there is a rich artisanal tradition of rendering the river of the kayva text in images. Artisanal production, unlike artistic production, occupies a very distinctive time and space as historical evidence, and new histories of craft read artisanal production as a site for alternative historical understanding.[44] Even though artisanal productions are not traditional pieces of evidence, to nonetheless ignore the rich repertoire of artisanal representative practices of rivers and riverbanks would be a serious limitation to recovering the human–land–water relations in the Bengal Delta. The

[42] For the theorization of the erasure of movements in mapping practices, see Carter, *Dark Writing*, 5–8.

[43] Lefebvre, *The Production of Space*, 45 and 31–53.

[44] Pamela H. Smith, Amy R. W. Meyers and Harold J. Cook, *Ways of Making and Knowing: The Material Culture of Empirical Knowledge* (Ann Arbor, MI: University of Michigan Press, 2014).

paintings of Bengal's richly imagined riverine space, as shown in Figure 3.1, which are still available as artisanal productions and not as state-historical archival remains, portray their world in a nonrealist representative index. These paintings are integral elements of the narratives that are sung. As a performance genre without the fixed temporality of a painting, the scrolls occupy a different aesthetic register, one that is not based on authorship or representational facticity. Their power lies in a metaphorical invocation of the spaces.[45]

Figure 3.1, depicting a scene from *Manasamangal*, lacks the controlled horizon of European landscape paintings and has none of the calm of the picturesque, as depicted in Figure 3.2.[46] The *Manasamangal* pictorial scroll (Figure 3.1) depicts the world of the vibrant human, animal and water–land relations that this space embodied.[47] The sacred geography of the ghat in the painting seems in no way threatened by an economic world order, but instead the sacred and economic are coterminous. Trade, worship and death existed along with humans, fish, plants and animals. The riverbanks depicted in the *Mangalkavya* texts are sites where goddesses appeared in human guises to trade and engage in material exchange. These scenes (Figures 3.1 and 3.3) bring forth a novel way of comprehending the space of the ghat, one that intimates other spatial relations, as well as heterogeneous ways of relating to the water, land and the space where they mix.

Figure 3.3, by an anonymous Indian artist in the early nineteenth century, illustrates the embedding of the vibrant market space in the sacred life of the ghat area. The absence of any clear demarcation of where the river ends and land begins is significant in this painting. For the painter both water and land are spaces of dwelling, selling and soaking.

[45] Discussing native American masks as ritual performances rather than as museum pieces, Gloria Anzaldua evokes the same argument about how to approach these objects outside a Cartesian understanding of them. Gloria Anzaldua, *Boderlands/La Frontera: The New Mestiza* (San Francisco, CA: Aunt Lute Books, 1987), see esp. 90–108; for recent scholarship on narrative scrolls and their social role, see Pika Ghosh, "Unrolling a Narrative Scroll: Artistic Practice and Identity in Late-Nineteenth-Century Bengal," *Journal of Asian Studies* 62, no. 3 (2003): 835–71; on using materials such as stones as an animate communicative objects, see Carolyn Dean, "Rock Sites/Rock's Sight: Reflections on Site Documentation," *Public Art Dialogue* 1, no. 2 (2011): 151–61.

[46] Tapati Guha-Thakurta, "The Compulsions of Visual Representation in Colonial India," in *Traces of India: Photography, Architecture, and the Politics of Representation, 1850–1950*, ed. Maria Antonella Pelizzari (New Haven, CT: Yale University Press, 2003), 14 and passim.

[47] For a recent analysis of the *Manasamangal* narrative, see Kaiser Haq, *The Triumph of the Snake Goddess* (Cambridge, MA: Harvard University Press, 2015).

FIGURE 3.1 Present-day pictorial scroll depicting scenes from the kavya *Manasamangal* (undated), From Medinipur, West Bengal.
© Wikimedia Commons. https://upload.wikimedia.org/wikipedia/commons/a/a7/Manasa_Mangal.jpg.

FIGURE 3.2 "View of Calcutta from the Garden Reach." William Daniell R.A. Plate 49.
© British Library Board WD139.

The folk imaginations of the space are in direct contrast to the representations of continuous and controlled vistas in colonial paintings (Figure 3.2) and cartography, where the coastlines and riverbanks arrange the coherence of the representative index.

The ebbs and flows of the river make and unmake the city, giving the city both its natural boundary and cultural form. Yet few histories pay adequate attention to the life-world of that space, a life-world not simply defined by a network of relations (be they economic, political or social), but rather characterized by thick routes of differentiation where both actual and potential inhabitation take place.[48] Riverbanks are places where deities, saints and weekly markets make claims and jostle for space without unsettling each other. The economic and sacred overlap to such a degree that any demarcation between them appears rather contingent and accidental.

Recovering this remembered landscape where the river and land form a continuous space of habitation is integral to rethinking the question of

[48] I use Bruno Latour's definition of life-world here. Bruno Latour, "Some Experiments in Art and Politics," e-flux Journal 23, March 2011, www.e-flux.com/journal/some-experiments-in-art-and-politics/, accessed 27 July 2013.

FIGURE 3.3 People bathing and praying in the Holy River Ganges, gouache painting on mica by an unnamed Indian artist. Date unknown. Image 15.8 × 19.8 cm. Courtesy of Wellcome Library, London,

ownership along the swampy banks of Calcutta.[49] Narratives, songs, paintings and storytelling are sites of claim-making that remain outside the proprietary gestures of mapping. They are sites to recover a varied range of spatial economies that maps and surveys render illegible.[50] In colonial Calcutta, the British preoccupation with land evaluation of the riverbank gradually eclipsed certain nonpropertized relations to land, although the latter did not disappear altogether. Rather, the sacred and

[49] Amanda Kearney and John J. Bradley, "Landscapes with Shadows of Once-Living People: The Kundawira Challenge," in *The Social Archeology of Australian Indigenous Societies*, ed. B. David, B. Baker and I. J. McNiven (Canberra: Aboriginal Studies Press, 2006), 182–203. Carter, *Dark Writing*, 54; see also Edney, *Mapping an Empire*.

[50] Denning, *Islands and Beaches*, 86–94. See Clifton Crais on how certain "facts" about vacant land legitimizing colonialism circulate through both visual and literal representational practices in "The Vacant Land: The Mythology of British Expansion in the Eastern Cape, South Africa," *Journal of Social History* 25, no. 2 (Winter 1991): 255–75.

profane cultures of the riverbank persisted in residual forms as illegality, squatting and fictitious claims, as "recalcitrant" forms of land usage, which ultimately brought the Hindu deity to the court.[51]

THE MANY BIOGRAPHIES OF ONE PLOT OF LAND

Property is multivalent and its social life is strangely opaque. There is no better place to explain this multivalent nature and the challenges it posed to the bureaucratic dictates of property regulations than the complex, layered, affect-saturated and mobile riverbanks. The multiple ways of understanding the river's relation to its banks were inconsequential to the European settlers. For the merchants of the Company and their trading interests, water and mud had to be separated and transformed into a recognizable geographical feature.[52] Separation of water and fortification of that land would bestow on Calcutta a bank, instead of the knee-deep swamp that linked (and continues to link) the land to the river. Moreover, throughout the early-modern and colonial periods, the banks of the river occupied an integral space in Calcutta as a port and a site of trade. The early English, Dutch and Portuguese factories around Calcutta were located along the river. The swampy and marshy riverbanks also figured significantly in the debates about sanitation, epidemiology and cholera throughout the nineteenth century.[53]

Traditionally, the spaces of the ghat were always open to access by the residents around that ghat, although use of the space did not confer any kind of proprietorial rights on the users. Due to its sacred nature, as well as putative usage for water and trade, the ghat served the purposes of public utility as a trading place, bathing place and small fruit and flower market that formed part of the temple economy around the ghat.[54]

[51] I employ the term "recalcitrant" in the sense used by Shahid Amin, *Event, Metaphor, History: Chauri Chaura 1922–1992* (Delhi: Oxford University Press, 1995), to mean events (or claims) that do not fit the given terms of a particular narrative or political articulation.

[52] A similar battle with mud and land took place over centuries in the Mississippi Valley, where the technological separation of land and water forms a critical moment in its history. Morris's history of the Mississippi Valley locates Hurricane Katrina as one of the consequences of what he refers to as the repression of the wetlands. See Morris, *The Big Muddy*.

[53] Smith, *Report on the Drainage and Conservancy of Calcutta*, 9–12.

[54] Wealthy merchants held markets on the ghats twice a week for a couple of hours. These small markets, known as *haats*, became a site of contention for the colonial officials. In 1855, after Strand Bank was declared a public space, the Chief Magistrate of Calcutta complained to the Secretary of the Judicial department about the rights of the wealthy

Wealthy Indian property owners with riverfront property would build steps leading to the water for the use of the public, not for monetary gain, but in order to enhance their prominence within the local community. It was common practice for landed proprietors, who occupied the second tier of power after the Mughal Governors in Bengal, to maintain law and order and undertake public works often in forms of ritual gifting (*dana*), alms, and sacrifice.[55] This custom explains why most ghats in the Gangetic plain are named after wealthy merchants and land proprietors.[56] As Swati Chattopadhyay argues, "[t]he ability to, pay, however, was not the only determinant of spatial use or social power."[57] Ghats were always considered shared spaces with their own distinct caste-class hierarchies and politics of ownership. It is not surprising that the Indian landowners not only fought bitterly about land acquisition along the ghat, but also resisted every intervention into reordering the bathing ghat, crematorium or ghat markets.[58]

The riverbanks encapsulated a sacred geography where ancestral spirits and gods resided, making the land invaluable and rendering multiple social, economic and cultural worlds possible. Wealth and political power had their own nonsecular logics within South Asia. We can locate the case of the Hindu deity, who thwarted the colonial expansion of

Indians to hold markets, since "In no other part of the city, that I am aware of, does such a right exist, & by the provisions of Act XII of 1852, the sale or exposure for sale of articles on the public street is prohibited and is punishable by offence." Compensation claimed by the Rajas of Shobha Bazar for Their Land on Strand Bank, Judicial [Judicial] 18 October 1855, Prog. 22–24, WBSA.

55 Numerous studies on the landed gentry of Bengal attest to this. See Shirin Akhtar, *The Role of Zamindars in Bengal, 1707–1772* (Dhaka: Asiatic Society of Bangladesh, 1982); Ratnalekha Ray, *Change in Bengal Agrarian Society 1760–1850* (Delhi: Manohar Publishers, 1979); John R. McLane, *Land and Local Kingship in Eighteenth Century Bengal* (Cambridge: Cambridge University Press, 1993); and Rajat Dutta, *Society, Economy and the Market: Commercialization in Rural Bengal, c. 1760–1800* (Delhi: Manohar Publishers, 2000).

56 Radharaman Mitra, *Kalikata Darpan*, 253–315, documents the history of each of the ghats along Calcutta, attesting to how wealthy families often performed charitable work by constructing the ghats and making them available for use to their tenants and neighbors living in that particular area.

57 Chattopadhyay, *Representing Calcutta*, 80.

58 The Judicial Proceedings from the second decade of the nineteenth century are full of petitions against the interventions of first Lottery Committee and later Justices of Peace in the ghat area. See Neemtolla Ghat, Judicial [Criminal] 31 August 1826, Prog. 22, WBSA, Kolkata; Disapproval of the Proposal of the Justices of Peace for Calcutta to Take More Land for the Purposes of Burning Ground at the Kossy Mitter's Ghat, Judicial [Judicial] March 1865, Prog. 18–22, WBSA, Kolkata; Acquisition of a Strip of Land in Jackson Ghat, Judicial [Judicial], February 1866, Prog. 49.

Calcutta's aquatic resource frontier, within this nonsecular logic. The bureaucratic state machine mustered several inventive legal categories to accommodate ancestral spirits and Hindu deities within the architecture of the legal regime of urban property.

The case of the Hindu deity shows how the colonial officials and indigenous property owners negotiated the uneven terrain that began to emerge between imperial governmentalities and market practices regulating landed property. Following the publication of the Strand Bank Report of 1820 analyzed in Chapter 2, the Company decided to acquire lands for the construction of Strand Road. The contentious process of land acquisition began from the latter half of the 1830s, although the construction of the Strand Bank as a public wharf did not begin until 1852. The period following the publication of the report was marked by a gradual process of acquisition that significantly changed the power structure of the urban land market. The judicial process of acquiring the accreted land as government land reveals the legal "realities" created by the Company's infrastructural architecture along the riverbank.

The Lottery Committee, which was formally responsible for town improvement since 1817, although informal forms of lottery sales for town improvement existed since 1793, began to acquire the lands for the Strand Road.[59] One of the first parcels of accreted land belonged to Raja Radhakanta Deb Bahadur, the scion of a wealthy Calcutta merchant. In 1838, the City Magistrate forced Deb Bahadur to forfeit his claim to what he referred to as his "river frontage" property. After repeated appeals failed to soften the Committee's stance on his land, he finally chose to give up his land to the Committee. The Committee promised compensation, though they never made good on their promise. His tenants and the people living around his house had long used the "newly formed alluvion," which the Lottery Committee claimed for the public wharf. The Company's acquisition reversed the longstanding communal use of the space. In his appeal he informed them of the great

[59] Using lottery sales to raise money for town improvements was practiced in both metropolitan and colonial towns. In Bengal, the practice began as early as 1793. Between 1805 to 1817 funds collected through lottery sales were given to Lord Wellesley's Town Improvement Committee, which funded the digging of wells, building of tanks, construction of the Town Hall and building of the Beleghata Canal. In 1817, the Lottery Committee was formally appointed and continued to function until 1836 when it was replaced by the Justices of Peace. For a detailed history of the lottery committee, see Binay Krishna Deb, *Early History and Growth of Calcutta* (Calcutta: Bengal Printing Company, 1905), 47–50; for a more recent account, see Datta, *Planning the City*, 18–28.

inconvenience it would cause to his tenants if their daily activities at the river were discontinued due to the construction of a public wharf.

The space between his house and the river was also a religious endowment held by him for communal use. "River frontage property" like Deb Bahadur's, which often consisted of temples or water reservoirs, was known as *devuttur sampatti*, a Hindu form of religious endowment.[60] According to Hindu law, owners could relinquish their rights to such property through a gift or a ritual act of renunciation to the public, while still retaining either absolute or partial control as a donor, not as an owner.[61] However, before the Hindu deity was reified as a legal entity through the Charitable Endowment Act of 1890, properties of this nature posed a legal conundrum.[62] Not surprisingly, Deb Bahadur's appeal went unheeded by the members of the Lottery Committee.[63] The Committee took over the land by using what Nasser Hussain describes as colonial law's "plenitude of prerogative" – its ability to justify its stance in the name of the general public welfare and the city's transport needs, especially once the Committee's "constitutional authorization [was] exhausted, with the claims of legal propriety spent."[64] Deb Bahadur's appeal was based on the contention that the people living in that area and his tenants were already using his land, and that this parcel was never held by him as private property. For the colonial state, however, such heterodox communal spaces were not necessarily public. The Company, in its new self-fashioning as a public agent of land, redefined public space as commercially productive and revenue generating for the municipal coffers. Such exceptions to the law produced and shaped the legal geography of the city. As the 1820 Report indicated, the tidal land between the high-water and low-water marks belonged to the "public," even if the adjacent riverbank was possessed in full and indisputable proprietary rights by any particular individual and used by their

[60] Bijan K. Mukherjea, *The Hindu Law of Religious and Charitable Trust* (Calcutta: Eastern Law House, 1970), see esp. chap. 7.

[61] "Sattva Vicara," in Derrett, *Essays in Classical and Modern Hindu Law*, 478–79. Some studies attribute authorship of this text to Raghunath Siromani; however, Derrett elaborates on the existing debates that deny sole authorship to Siromani. See also J. Duncan M. Derrett, "The Administration of Hindu Law by the British," *Comparative Studies in Society and History* 4, no. 1 (November 1961): 10–52.

[62] See Birla's analysis on the reification of the Hindu deity in *Stages of Capital*, 87.

[63] Radhakanta Deb versus Company Government, Judicial [Criminal], Letter to George Earl of Auckland, Governor of Bengal, dated 10 November 1840, Prog. 54, WBSA, Kolkata.

[64] Hussain, *Jurisprudence of Emergency*, 2.

tenants.[65] The appearance of the Hindu deity and its claim put to test the Company's collapsing of communally used space into state-controlled public space.

In the following case that I discuss, the colonial power had to confront the sacred authority of the Hindu deity when case of the wealthy shipping merchant Moti Lal Seal (hereafter the Seal case), came to the Privy Council. In 1851, about a year before the construction of the Strand Bank, the government began the process of acquiring another parcel of the newly formed alluvial land in front of Seal's property.[66] Seal, who was thoroughly enmeshed in the mercantile machinery fueled by British capital, handed over the land to the Company agents without much protest. However, he stipulated two special provisions. The first provision was that if, for the purposes of improving the land, the government felt the need to demolish the landing and bathing ghat that existed there, it would be obligated to construct a new one. The second provision stated that the land could not be used for other purposes apart from building a road.[67]

By handing over his rights in this manner, Seal became a tenant on his own property and continued to pay a monthly rent to the government until it was eventually taken over for "public purposes." When in 1857 the Lottery Committee finally tried to bring the land into its infrastructural grip, Seal's eldest son and heir protested. Subsequently, two separate legal battles ensued. These battles were characterized by the multiple articulations about rights and equity that moved through the various city courts, the Privy Council and finally the High Court in Calcutta. The Privy Council was often the highest institution for appeal in a dispute involving Indian petitioners and the East India Company; however, one of Seal's two cases went beyond the purview of the Privy Council. One important development took place between 1851 and 1857, during the time between Seal's relinquishment of his rights to that land and his son's petition to the Privy Council, a change attesting to the altered status of the Strand Bank in official nomenclature. In October 1854, the Chief Magistrate of Calcutta declared the riverbank

[65] Extract from the Proceedings of the Territories Department, 1 April 1820, Prog. 15, §51, WBSA, Kolkata.

[66] For more on Moti Lal Seal, see Sabyasachi Bhattacharya, "Traders and Trades in Old Calcutta," in *Calcutta, the Living City*, vol. 1, ed. Sukanta Chaudhuri (Calcutta: Oxford University Press, 1990), 209; Kissory Chand Mitra, *Mutty Lall Seal* (Calcutta: D'Rozario Printer, 1869).

[67] Legal Tussle with Strand Bank, Judicial [Judicial], 1 June 1870, Prog. 161, WBSA, Kolkata.

to be a public space, prohibiting all forms of communal activities apart from commercial ones on the site.[68]

Small temples and huts fell under the ambit of this prohibitive order. However, removing and harnessing the recalcitrant Hindu deity on Seal's land proved difficult for the Company, for this was not the only case of a deity making legal claims to land. Within a pantheistic Hindu context, these idols occasionally appeared "surreptitiously" in various kinds of spaces, especially in spaces of public works projects, to the great consternation of colonial officials. The Seal case was one among the many cases that necessitated the reinvention of Hindu deities as legal subjects. Like the Deb Bahadur case, this episode took place nearly thirty years before the passing of the Charitable Endowment Act of 1890, which transformed the legal status of Hindu deities.[69]

In 1857, when the government took possession of Seal's land, they nullified his son's petition and asked him to remove everything, including, as the memorandum stated, "various idols and objects of Hindu worship." Despite these efforts, the Judicial Proceedings of 1870 reports the following incident on Seal's land: In April 1864, "a small idol was surreptitiously set up under a tree standing in the centre of the Government Land, and a rude mat hut erected over it, in which sat daily a Brahmin priest, whose ostensible function it was to mark the foreheads of the bathers."[70] This "surreptitious" occupation of land continued for six years. Finally, in 1870 the government sought to remove the idol and "a claim was set up to the possession of the land on which the idol stood, on the *ground of a fictitious occupancy*."[71] The Brahmin priest (*shebait*) did not claim the space in terms of absolute ownership; however, his followers gave his deity offerings, which was also his livelihood. By taking the land away from the deity, the government not only denied the priest rights to his livelihood, but also denied the devotees their rights to the river. After a protracted

68 Declaration of Strand Bank as Public Space, Judicial [Judicial], 19 April 1855, Prog. 69–73, WBSA, Kolkata.

69 From 1870 to 1890 earlier attributes of Hindu gifting and philanthropy were substantially remapped within the legal sphere. In 1890, Hindu deities were transformed into taxable entities in order to control and separate the traffic between the secular and commercial worlds. For a detailed reading of the Charitable Endowment Act and the separation of the sphere of religious from economic practices, see Birla, *Stages of Capital*, 96–102.

70 Legal Tussle with Strand Bank, Judicial [Judicial], June 1870, Prog. 161, §2, WBSA, Kolkata.

71 Legal Tussle with Strand Bank, Judicial [Judicial], June 1870, Prog. 161, WBSA, Kolkata (emphasis mine).

legal battle with the deity, the government was forced to pay compensation for the relocation of the "temple" or the makeshift hut and its shebait before the land could be made available for Strand Bank purposes. It must be noted that this was the only compensation the government paid to the former owners of land acquired along the Strand Bank.

As we know, the English maintained a strict separation between the religious and secular orders. Warren Hastings, the first Governor-General of India, introduced this separation to the colony in the process of codifying the Indian legal system in 1772 by bifurcating the arenas of civil law. He stipulated that, within the new legal codification, the common and natural law doctrines of the British legal system were to be used to adjudicate territorial cases through procedural and substantive law.[72] In matters of religion, which colonial officials classed under personal law, pertaining broadly to family, caste and religion, Hastings deemed that religious affiliation would determine whether Indians were governed by Hindu or Islamic laws. By determining family and community matters through the dictates of religious laws, the colonial state could maintain its pretense of noninterference.[73] Yet the liberal legal doctrine of property, which was grounded on British law and the common-law interpretation of Indian customs, conflicted with the sphere of religious laws, wherein property was donated to Hindu gods and could be held by them in perpetuity.[74] The Seal case collapsed the religious and secular divisions and broke down the modern language of contractual property claims that the colonial lawyers sought to ascribe to the deity's land.[75]

[72] Sturman, *Government of Social Life*, 6–8.

[73] Chatterjee sees this as limits of the modernizing impetus of colonial legal architecture and argues how such practices then became sites to attest colonial difference to universal legal principles. Chatterjee, *Nation and Its Fragments*, 14–33. For scholars who have looked at how this dual structure shaped colonial India, see Bernard S. Cohn, "Law and the Colonial State in India," in *Colonialism and Its Forms of Knowledge: The British in India* (Princeton, NJ: Princeton University Press, 1996), 57–75; J. Duncan M. Derrett, *Religion, Law and the State in India* (Delhi: Oxford University Press, 1999); Marc Galanter and Rajeev Dhavan, *Law and Society in Modern India* (Delhi: Oxford University Press, 1989).

[74] Such forms of possession should not be confused with mortmain or trust properties, which had intricate systems of ruling and adjudication, especially in Islamic and Parsi legal cultures. Leilah Veviana, "Trust Matters: Parsi Religious Endowments in Bombay-Mumbai" (unpublished manuscript), addresses the creation and management of Parsi endowment properties from the British courts to the present moment.

[75] Waqf properties (charitable endowments in Islamic legal tradition) charted a slightly different trajectory; see Beverley, "Property, Authority and Personal Law: Waqf in Colonial South Asia," 155–82.

Thus, instead of reading this as a case of strategic resistance to infrastructural projects, I read it as a site where the modernizing impetus of colonial property law breaks down.

The liberal and secular nature of colonial ideas pertaining to property in riverine land was rendered inadequate to the issue at hand. Precisely because the colonial legal logic broke down in the face of the Hindu deity, unlike in Deb Bahadur's case discussed earlier, the Company did pay compensation, however trifling, to relocate the priest and his deity. This case reveals the inconsistencies inherent in colonial laws of property that sought to demarcate spaces through fixed lines and enforce contracts through bounded categories.[76] The secular stance of colonial law succumbed to the divine logic of the riverine deity, if not the hydro-logic of the landscape.

Further, Seal's family was in possession of this newly formed alluvion continuously for twelve years, and thus, according to Regulation XI of 1825 pertaining to alluvion's adverse possession, this piece of land was legally theirs to claim. They had the right to apply for a title to turn a piece of benami sedimentation into their property. However, Seal's son chose not to do so. During his deposition to the Company, he mentions that part of the land where the deity cropped up was religious land. However, since the land was of a "trifling nature" and was used by the people, he chose not to infringe on their daily practices and access to the river, unlike the British, who left no stone unturned in the process of eviction.[77]

Seal's son and Indian landowners in similar situations did not necessarily regard such religious "occupancy" of their land as a breach of personal property, especially given that the property was linked to a devuttur, or donated property. We do not know whether the priest or the deity fought the case at Seal's behest, or whether allowing the Brahmin to sit with his deity was an elaborate ploy by Seal to frustrate the Company's attempts at acquisition. The two cases, the case by Seal's heir and the Hindu deity, were filed at two different courts and these papers are currently housed in two different archives.[78] The Company's agreement to pay compensation and to succumb to the demands of the deity, even when its occupancy was deemed "fictitious," reveals the ambivalent

[76] Similar issues of incompatibility in governing personal lives through law has been explored in detail by Sturman, *The Government of Social Life*, 27–29, 70–106.

[77] "Mutty Lal Seal's Bathing Ghat, Secretary's Department, Calcutta Port Trust," 929 PT (Kolkata, Kolkata Port Trust Archive), hereafter Mutty Lal Seal 929 PT.

[78] While Seal's papers are at Kolkata Part Trust Archives, the case of the Hindu deity is archived with the Judicial Proceedings of the West Bengal State Archives.

legacy and the limited nature of colonial legal architecture on riverine land. These limits came into relief in a space where the local inhabitants believe that the river, which gives the land, has the highest claims on the land.[79] To bracket off the deity's claim as merely a strategic move to resist the infrastructural expansion and the disciplinary regime of the colonial state is to obfuscate the multiple meanings that abound in these silty riverine spaces between dry land and flowing water.

However, the case by the Hindu deity was not the only one that was fought on Seal's land. The other legal case was neither resolved through compensation nor settled until 1899. Unlike the case of the deity, this case reveals a different texture in the legal geography of the city. The Company acquired Seal's land through a Bengali *kubhoobut* (paper granting land signed by two parties) in 1851 that was supposed to be in the possession of the government of Bengal but subsequently went missing.[80] This piece of paper had a few stipulations, first that the government was permitted take possession only when it was ready to begin work, second that it would construct a new landing area if it tore down the existing one, and third that the land could not be used for any other purposes except the construction of the Strand Road. After misplacing the paper, the government ordered the Calcutta Police in 1852 to prepare another "instrument in writing," stipulating the terms of Seal's land grant to the government of Bengal.[81] The Calcutta Police altered some of the provisions in this new "instrument in writing," and these alterations became one of the flashpoints in the conflict that ensued.

This "instrument in writing" made Seal a tenant in his own lands, for which he paid a monthly rent of 10 rupees. When Seal passed away in 1854, his heirs inherited this land.[82] In 1856, Samuel Wauchope, who had by then become the Chief Magistrate of Calcutta, ordered Seal's heirs to vacate the land, which they refused to do. After two notices Wauchope filed a case against Seal's inheritors – three sons and one daughter. The court ruled in Wauchope's favor.[83] During the court case the Seal family

79 Nineteenth-century legal discussions pertaining to riverine ownership often referred back to an Indic source by Vrihaspati, who purportedly stipulated that a river, like the sovereign, had the power to gift and take land away. See Lal Mohun Doss, *The Law of Riparian Rights, Alluvion and Fishery: With Introductory Lectures on the Rights of Littoral States over the Open Sea, Territorial Waters, Bays, &c., and the Rights of the Crown and the Littoral Proprietors Respectively over the Fore-Shore of the Sea*, Tagore Law Lectures (Calcutta: Thacker, Spink and Co., 1891), 178.

80 Mutty Lal Seal 929 PT: 15.

81 Mutty Lal Seal 929 PT: 16.

82 Mutty Lal Seal 929 PT: 18.

83 Mutty Lal Seal 929 PT: 20.

discovered that the police, in drawing up a copy of the kubhoobut, had altered the provision from "road construction," as Mutty Lal Seal had initially stipulated, to "any kind of general improvement." These revisions by the Calcutta Police made it possible for the Chief Magistrate to acquire the land and to ignore earlier provisions. The Company successfully applied the "plenitude of prerogative" that is vested in the all-encompassing clause of "any kind of general improvement."[84]

The Seal family did not give up on their claims easily and in 1857 they filed for a retrial in the Privy Council. The Privy Council ruled in favor of the earlier judgment by the court in 1856, stating that the "instrument in writing in 1852 had merely created a tenancy at will in Mutty Lal Seal."[85] However, the Privy Council also added that the legal interest in the land had been passed to Mutty Lal Seal through the aforementioned instrument in writing and that Seal's equitable rights in land could only be addressed at the Court of Equity.[86] Although the question of equity often came up in Privy Council cases, what the British understood by equity was never properly defined, and equity remained one of those open-ended words within the legal world, which allowed for many possible interpretations.[87]

The Seals took the case to the Court of Equity in their bid to make the British fight for this parcel of land. This case eventually moved to the High Court of Calcutta and was settled in 1869. From 1852 to 1869 the nature of the case completely transformed. While the High Court ruled that the Seals would have to give up the land, it also ordered the government of Bengal to rebuild a masonry landing area, in case they had to demolish the existing one, which had been built by Mutty Lal Seal.[88] While the Court granted the wishes of the Seals, it also ordered them to pay for more than half the cost of constructing this new landing area and for any other ornamental decorations.[89]

While the government and the Seals fought over all the minute details surrounding the construction including the placement of trees and the nature of the road, two disputes stand out in particular. The first one was regarding an apostrophe and the second one dealt with what the plaintiffs termed the "space of mud." When the final structures of the landing area

84 Hussain, *Jurisprudence of Emergency*, 2 and 21.
85 Mutty Lal Seal 929 PT: 22–23.
86 Mutty Lal Seal 929 PT: 24.
87 For a discussion on the idea of equity in colonial legal thinking, see Chatterjee and Subramanian, "Law and the Spaces of Empire."
88 Mutty Lal Seal 929 PT: 36.
89 Mutty Lal Seal 929 PT: 37–38.

were completed in 1875 with funding from the Seals (the government only paid to demolish the existing landing area and fortify the alluvion), the government declared it "Mutty Lal Seal Public Ghat." The Seals objected to that wording and demanded that the government name it Mutty Lal Seal's Public Ghat, a claim made in accordance with the provisions regulating charitable donations in Hindu laws.[90] The government retorted that placing the possessive ending might create private rights in a public property and it refused to accept the reformulated dedication of the landing area.

While the colonial state might have won the apostrophe battle, the specific ecology of the silt became another tool of resistance in the hands of the Seal family. Seal's daughter Anandamoyee Dasi, who was also a plaintiff in the earlier cases, filed a separate petition regarding the ever-expanding "space of mud" on the landing area.[91] Rallying around the annual silt deposition that choked the newly constructed passage and steps to the river, Anandamoyee sought out a method to frustrate the colonial officials who now claimed public ownership of the space.

In its cartographic-mindedness, the colonial state had fortified a moment in the movement of the river. That hardly meant that the river would not move again. In keeping with the tidal behavior of the river, the Hooghly River started moving and depositing new silt on the landing area, creating the "space of mud" that bathers had to traverse to get to the river. The fortification of the space, the draining of the sediments for Strand Road, the construction of the new ghat and attempts to commercialize the space were being engulfed by the onslaught of new silt and mud. The British were beginning to realize that maintaining the hydrological infrastructure in this city would be an ongoing and expensive affair. The fight began to determine who was responsible for removing the silt and maintaining access to the river. Having given the land to the government, the Seals claimed that they also transferred the responsibility for the maintenance of the riverbank to the city officials. By 1890 the mud was so deep and thick that it created a danger of drowning. Finally, in 1899 the Port Trust removed the silt and stipulated that it was the Port Authority's duty to remove it once and slope the sides, after which people with riverfront property would be required to maintain clear access to the river.[92] Once again, the western banks of Calcutta were submerged in

[90] Mutty Lal Seal 929 PT: 253–54 and 298.
[91] Mutty Lal Seal 929 PT: 154–75.
[92] Mutty Lal Seal 929 PT: 356–63.

knee-deep mud from which the British had sought to rescue it since the publication of the Strand Bank Report in 1820.

INFRASTRUCTURES OF POWER

The area along the river, both a market and a sacred space, was slowly and contentiously transformed into property. The Deb Bahadur and Seal cases are just two among the many petitions from Indian property owners along the river and in various other parts of northern Calcutta that fell under the intense spatial expansion of mid-nineteenth-century urbanization.[93] These cases along the riverbank also illustrate the maturation of the quasi-eminent domain articulations within the hydrological fixing of the urban marsh that began in 1820. There were many reasons for mobilizing the 1820 report's stipulations for acquiring these plots of land. The lands adjoining the Deb Bahadur and Seal properties or even the land where "a Brahmin priest daily sat" were used by people living in the area and these usages have a long history. Making land, especially along riverbanks, available communally through gifting practices was one of the ways of creating and sustaining political authority in precolonial times. For instance, Indic kingly authority, from as early as the eleventh century, was sustained through generous and symbolic gifting practices in the form of temples, water reservoirs and centers for learning, which enabled a translation of material resources into status and authority.[94] Broadly speaking, we witness something similar during the Mughal period in the Indo-Gangetic plain, where senior-level functionaries funded infrastructural projects, especially canal and road construction. These projects were exempted from taxation, generated profits and conferred symbolic status.[95] Such practices persisted even after the arrival of the British and were accommodated into their modernizing tendencies in different ways.

If Lacam attempted to fashion himself as a merchant-speculator-cum-Mughal-functionary by coopting practices of infrastructure financing

[93] Mallicks versus Company Government, Judicial [Criminal], 4 September 1818, Prog. 6–10, WBSA, Kolkata; Petition to Save the Sanctity of Religious Spaces, Judicial [Criminal], 22 August 1822, Prog. 35–41, WBSA, Kolkata.

[94] Kingly authority was sustained by elaborate networks of temples, especially in Southern India. See for instance Arjun Appadurai, "Kings, Sects and Temples in South India, 1350–1700 A.D.," *Indian Economic & Social History Review* 14, no. 1 (1977): 49–73.

[95] Hameeda Khatoon Naqvi, *Urbanisation and Urban Centres under the Great Mughals 1556–1707: An Essay in Interpretation*, vol. 1 (Simla: Indian Institute of Advanced Study, 1972), 18–19.

with the hope of monetary profits and social mobility, Deb Bahadur and Seal were consolidating their own social authority within the nineteenth-century Indian urban milieu by donating their land for sacred purposes. And what could be a better site for such consolidation than the rich deposits of the holy Hooghly River?

The practices of creating social authority within the emergent colonial regime were eroding under the centralized bureaucracy of the modern state, where town infrastructure, planning, expansion of communication networks and policing increasingly became sites of state political control. To this list, we can now add economic governance exercised through modes of propertizing. Hence, instead of reading these contestations merely as resistance to the colonial state's public works projects, this chapter located them as sites where one can cipher the changing nature of authority surrounding land and water through the nineteenth century. These cases signal a shift away from embedding social and cultural authorities into spatial practices to a process of monetizing spaces by the colonial state. Riverine movements and the specific littoral ecology played a central role in the development of this particular territorial consciousness. The everyday lived histories of the contestations and surveys documented here show the widening gap between the imperial regime of property and the lived hydrological histories of places imbued with multiple meanings. The questions pursued in the following chapters involve investigating how projects of restructuring landholding in muddy silt impacted the emerging land market in colonial Calcutta.

4

Commerce in Land

> One cannot get far with ... town-planning ... without knowing the behavior of the matrix in which the foundations of the work are to be set, and this is surely the case in a country like Bengal where the face of the earth is more or less in the state of flux.
>
> C. Strickland, *Deltaic Formation*, 1940

For much of the early nineteenth century the urban land market led an inconspicuous life with little connection to the infrastructural regime of the empire. However, this began to change from the second half of the nineteenth century. The nineteenth-century archive of urban land and property in Calcutta brims with multiple genres of petitions and court cases pertaining to ownership disputes. By the middle decades of the nineteenth century, infrastructure and the urban property market started emerging as intertwined entities and objects of municipal concern. This chapter begins in the middle decades of the nineteenth century when the political economy of urban property and the land market was gradually rearranged in a new manner through the infrastructural projects of the empire. What began as ad hoc attempts at producing land out of clay, silt and marshes for the protection of private property and the Company's landholdings gradually entered new domains of economic, social and political values. The concerns of profit that drove Benjamin Lacam's venture and the expectations of pecuniary gains from the Strand Bank project were now yoked in an expansive economic and new spatial logic. Acting as an intermezzo, this chapter connects the hydrological history of the city to the larger economic and legal forces unleashed by colonial rule.

A techno-bureaucratic regime of draining began to create new forms of land valuation in Calcutta from the 1840s onward, laying the groundwork for the emergence of a radically different urban economic geography. By the 1860s, extensive parts of the delta had been drained and legally zoned to create dry docks, wet docks, urban parks, canals and riverfront properties.[1] These changes elicit two questions. First, through what (licit and illicit) financial means did the colonial officials fund these massive projects of draining the city? Second, did these changes produce a new political economy of land, as substantial parts of Calcutta came under direct Company and later state ownership? At the heart of this chapter is an effort to show how the techno-logic of draining (and planning) cultivated amnesia about the soaking ecologies of Bengal. This amnesia created the conditions for the emergence of speculative abstraction of value from land in Calcutta at the turn of the twentieth century. This was also a period when legal experimentation gradually gave way to the instrumentalization of risk through new market practices and new ideas about land finance.

Three major draining projects fixed the cartography of Calcutta and reconfigured the relation between land, water, property and the urban land market beginning in the third decade of the nineteenth century. One of the projects was to reconfigure the southern part of the Strand Road into an extensively fitted harbor. Part of this project involved restaging earlier forms of private ownership of canals into the wider networks of Company and later state landholdings. The other project was to permanently drain the Maidan, 333 acres of open green space in the heart of the city. Alongside the projects of draining the swamps and transforming them into productive land, the Lottery Committee and the Sanitary Department laid down an extensive network of canals through and around Calcutta beginning a process of propertizing the water that existed within the marshy space. Each of these projects became a massive source of revenue. Some of them began as neo-mercantilist endeavors, yet ended as state-centric infrastructural models, where questions of profit were enveloped in a concern to rule. For instance, the canals of Calcutta were primarily used for navigation and communication with eastern Bengal, the Sundarban area and much of the hinterlands of the city. Most of these were canals dug along the natural drainage lines of the river

[1] Dutta documents this period as one of intense growth of urban properties and extensive public works, *Planning the City*, 127–70.

during the early modern period.[2] The canals in Calcutta were anywhere between 47 and 350 miles long, and by the final decades of the nineteenth century they reaped a handsome profit for the state. According to the estimates of the 1895–96 report, the capital expenditure on the canals of Calcutta was 5,486,353 rupees with a profit of 13,825,353 rupees.[3] Where did these stupendous amounts of money and profit go?

NEW ECONOMIES OF INFRASTRUCTURE

This history began at a moment when the Company had successfully fashioned itself as a Company-State and became one of the early European joint-stock companies to embody military-fiscal power with territorial holdings outside Europe. It was also increasingly becoming an international actor with legal rights and obligations.[4] In its early days, petty bureaucrats and merchants like Lacam could carry out infrastructural projects as private profit-making ventures with the Company's support. The making of the Strand Bank in Calcutta had a long gestation and execution period, beginning under the aegis of the Company and ending after the transfer of power to the Crown and the consolidation of the British Raj in India. As a research site, the Strand Bank helped us recover the local dimensions and impacts of the broad changes in colonial power.

As Chapter 2 demonstrated, the Strand Bank began as a project of managing alluvion in a tidal river and transformed into one of the centerpieces of the colonial urban infrastructural projects in Calcutta. The function of this project was manifold. It sought to discipline and displace the population living and working along the riverbanks, lay the

[2] It is hard to estimate the exact dates. Some of these canal routes still exist either in the memory of the people or sedimented in the names of the places. Many of these canals connect present-day Kolkata to cities on the current borders with Bangladesh. For instance, the Eastern Canals extended from the eastern tip of Calcutta in Dhappa to Barisal in present Bangaldesh and consisted of a series of artificial cuts that connect the natural drainage lines of the Ganga Delta. Bengal Irrigation Manual, vol. 1, 1897, IOR/W 559: iii–iv. Among the present dock workers of Kolkata there has been an effort to revive and document the extensive canal networks of Bengal. See Laura Bear, *Navigating Austerity: Currents of Debt along a South Asian River* (Stanford, CA: Stanford University Press, 2015), 3.

[3] Bengal Irrigation Manual, iii–iv.

[4] Philip J. Stern, "The English East India Company and the Modern Corporation: Legacies, Lessons, and Limitations," *Seattle University Law Review* 39, no. 2 (2016): 423–44.

foundations for grabbing land and create a spectacle of colonial infrastructure.[5] The Strand Bank also served as the backdrop for discussions of other projects related to managing water within the city, with an eye toward economizing space. The projects under discussion in this chapter include the privately funded creation of Tolly's Nullah, a canal that remained a source of revenue until 1893; the construction of dry docks in Khidderpore lying south of the Strand Road; and the expansive draining of the Maidan. What do these projects, spanning from 1840 to the final decades of the nineteenth century, tell us about the relation between the new technologies of imperial power and the management of water within the urbanizing landscape?

The changes in the expanding land market were linked to infrastructural expansion and the economic transformations within the empire itself. From the middle of the nineteenth century on, the fiscal policies of the empire evolved in response to the political developments in colonial India and global economic processes. In 1860, the India Income Tax Act fiscally restructured the urban economic geography of the British Raj with the introduction of income taxes, taxes on water, port taxes, property taxes and charitable property taxes. In linking the hydrological history of the city's expansion to its new economic geography, we can investigate how new circuits of value began to consolidate and restructure the already changing urban land market. New instruments of taxation and regularization of urban landholding constituted experiments in managing the ecological spaces not just through law, but also through new forms of market governance that facilitated a particular kind of growth in urban developmental projects.

Entrepreneurial merchants undertaking large-scale public works for future profit were slowly replaced by state involvement. Documenting this transformation in the production of an infrastructural landscape helps recover the emerging political and economic logic of colonial spatial practices, and demonstrates how these practices impacted the ecology of the city. New economic ideas and motives for rearranging the relation between urban land and water from the 1840s onward began to slowly supplant the idea of individual profit. The colonial coffers needed the voluntary and coerced contributions from the Indian financiers like Seal for their expansive town-planning projects. Nonetheless, they could not

[5] In his analysis of the World Exhibitions, Timothy Mitchell discusses the production of the colonial spectacle as an organizing principle of colonial power. Timothy Mitchell, *Colonising Egypt* (Berkeley, CA: University of California Press, 1991), 19–38.

afford to let their endowments create pockets of indigenous social and political authority. The Company would have to devise new plans to retain the private financing, fortify the city against the natural movements of the water, centralize infrastructural projects and benefit from the expanding land market.

BUILDING ON "BOTTOMLESS ALLUVION": WET DOCKS IN A DRYING DELTA

In June 1842, a massive hurricane struck Calcutta, resulting in a loss of property of the ships docked in Calcutta. The Company lost three ships within a day: One ship on its way to Calcutta was lost in the sands in Ghusuri, just above the city, and two government steamers sank. The Bengal Army Engineer W. N. Forbes responded by setting up an inquiry committee into the loss of life and property following this storm. He tasked the committee with submitting a report recommending measures that might be taken "[i]n the Port of Calcutta against the effects of similar gales, and against the dangers to which they are constantly exposed whilst lying in the River abreast of the city."[6] The committee, very much like Lacam's response to the mobile landscape, proposed that the government of Bengal work with the Marine Board to create wet docks in Calcutta. Fires and gales were of course a common occurrence, or, as one member put it, they were general losses one suffered in riverine trade, "but in the Hooghly they are exposed to such accidents in a greater degree perhaps than on any other river in the world, from the violence of the Bores at one season, and the strength of the Freshes in the other."[7]

Climate adaptation, one might conjecture, began in the early decades of the nineteenth century.[8] The Committee offered infrastructure as a response to the environmental vagaries and the "anomaly" that constituted the Hooghly River. However, unlike Lacam's project, these projects could not proceed without an initial calculation of revenue and profit for

[6] The Report of the Committee Appointed by the Government in May 1844 to Inquire into the Practicability of Providing Calcutta with Wet Docks, 1846, IOR V/26/750/2 (British Library, UK), 19 (hereafter Wet Docks IOR V/26/750/2).

[7] Wet Docks IOR V/26/750/2.

[8] This is not necessarily a far-fetched claim, as there were a few dissenting voices to construction projects close to the tidal river in an area prone to cyclones. See, for instance, how during the same period, Henry Piddington, who forecasted the problems with Port Canning, south of Calcutta, was ignored by the bureaucrats and builders, although he would be proved correct in a few years. Ghosh, *The Great Derangement*, 57–58.

the Company. One of the first tasks of the Wet Docks Committee was to estimate profits from taxes, tolls and duties on English, Indian and foreign ships. Apart from that the Committee was also tasked with ascertaining the depth of the water in the river and the suitability of the site. The Committee was confronted with two major questions: how to fund the project, and how to forge profits for a joint-stock company with only revenue rights. The Secretary to the government of Bengal wondered "whether the object in view [i.e. wet docks] could be effected by means of a Joint Stock Company, in which the Government would be considerable shareholder, as in the Bank of Bengal, or in what other way the Scheme could be most effectually carried out."[9] What, in short, would be the nature of the Company's investment, its role and estimated profits from this venture?

Scholarship on colonial urbanism has approached the nineteenth century in two ways. Some scholars argued miasma and epidemiology shaped early colonial urban planning. In contrast, others working on the post-1857 period studied imperial spatial practices and territorial consolidation as the emergence of a disciplinary regime.[10] This bifurcated focus sometimes creates an overly neat division between Company and Crown rule. For instance, in the case of epidemiological histories of the city, Chattopadhyay argues in *Representing Calcutta* that malaria and other diseases were the basis for a colonial pathologization of the Calcutta space. This resulted in the ascendency of the "environmentalist paradigm of disease" and the "climatic difference" between India and Europe as "central in understanding India's environment."[11] Overwhelming stress on epidemiology as a site for the production of colonial difference or the "uncanny" has obfuscated the market-driven aspects of infrastructure, and how they in turn created forms of market governance.

On the other hand, scholarship on intensification of infrastructure and urban spatialization of particular modes of colonial power, tends to identify 1857 as the moment of transition when two hundred years of Company rule ended and the British Crown took over.[12] The year 1857,

[9] Wet Docks IOR V/26/750/2, 4.

[10] Oldenburg, *The Making of Colonial Lucknow*; see also Legg, *Spaces of Colonialism*.

[11] Chattopadhyay, *Representing Calcutta*, 62–63.

[12] Veena Oldenburg demonstrates that Lucknow was severely affected and reorganized following 1857. Oldenburg, *The Making of Colonial Lucknow*. Following Oldenburg, many scholars identify post-1857 as a moment of intensification of infrastructural work. While Manu Goswami takes a long-nineteenth-century lens to study the shaping of infrastructural motives and ideologies, the focus of her work remains squarely in the

therefore, emerges as a watershed moment in this particular line of thinking. However, as the questions raised by the Wet Docks Committee clearly reveal, there was a distinct link between the hydrological urbanization of the city and economizing urban spaces, even prior to the consolidation of the "colonial-state-space," as indicated in Manu Goswami's work on the economic geography of India.[13] Accordingly, the disciplinary paradigm to study urban change in the post-1857 period, which was true in north Indian cities like Lucknow and Delhi, tends to obscure the longer genealogies shaping the logistics of colonial rule elsewhere. The miasmatic paradigm also ignores how colonial infrastructure was entangled in ecological interventions into nature and economic ordering of spaces.

Multiple failed iterations of expanding the dock infrastructure laid the groundwork for the emergence of a centralized project, based on new commerce in and from the lands of Calcutta. The 1842 project was not the first attempt at constructing a wet dock in Calcutta. However, this was the first one to be successfully executed under the aegis of the Company. Lacam's failed project was followed by General Henry Watson's project to construct a wet dock in Calcutta, which was aborted in 1824. Early infrastructure projects in the colony were privately funded, although the Company did support them with land and water grants. Watson's project, which was supposed to be privately funded, ran into the problem of government backing through grants. One of the early nineteenth-century cartographers and surveyors of Calcutta, Major Schlach, drew up a plan for Watson's proposed dock, which required him to occupy the Tolly's Nullah canal, privately owned by Major Tolly. Major Tolly bitterly objected to the plan through petitions and counter-petitions. When two of the important British members of the commercial community of Calcutta, Mr. Tolly and Mr. Watson, started bickering over the waters of Tolly Nullah, the Mercantile Members (constituted by some of the powerful members representing the merchant community in the Port Commission) interjected. They halted the project by declaring that they could not ascertain the extent of profit to be accrued from such a construction. If Schlach was concerned that the movement of water in the canal of Tolly's Nullah would be obstructed, then the economic future of

spectacular projects of railways, communication and irrigation, and not on the everyday maneuvers. See for instance Manu Goswami, *Producing India: From Colonial Economy to National Space* (Chicago, IL: University of Chicago Press, 2004).

13 Goswami, *Producing India*, 34–40.

the proposed docks was also under question. Watson's project was lost in a sea of petitions in the Mayor's court and was buried under debates around questions of state rights and the limits of private property, similar to the ones raging around the Strand Road.[14]

Three years after Watson's plan died, another Bengal civil servant named John Russell Colvin planned to build a dock in the northern part of the city between Meer Bahar (Meer Bhur) Ghat and Nimtollah Ghat from 1826 to 1827. Unlike Watson, he managed to secure permission to use the land and water adjoining the two ghats. However, this project also met an untimely death. The Anglo-Burmese War (1824–26) made any conversations around dock infrastructure impossible for the next twenty years. Finally, the hurricane of 1842 renewed discussions that finally materialized into wet docks in the Khidderpore area of Calcutta. Unlike Watson and Colvin's proposed projects, this dock was conceived and executed by the government of Bengal and was not a private venture.

The plans for the docks underwent multiple iterations. The initial docks were proposed to be as near to Tolly's Nullah as possible; however, the lands around the neighborhood were becoming increasingly expensive and there were numerous valuable houses in that area. The ground there was valued at 1,100 rupees per bigha, which would make the plan to build the docks in that area too expensive for the Company to bear.[15] In order to construct a dock capable of holding two hundred ships of four hundred tons in average weight, the Company needed thirty acres of water surface and twenty acres of land, which would include area for landing food sheds and a boundary wall. Given the staggering expense of acquiring land and constructing the docks, the Committee floated a new proposal. The committee decided that wet docks could be constructed cheaply if they were located further south by Akra and Garden Reach, where land was inexpensive in comparison to Calcutta, where most of the British residents lived. While the Acra plan was tempting, as far as land was concerned, this would require the extension of the railway lines along the marshy space in the southern fringes of the city. Prior experience already proved that such ventures were risky and expensive. Therefore,

[14] A letter from Lieutenant Colonel Henry Watson intimated that the Governor and Council of Bengal had entered into an engagement with Major Tolly that threatened to ruin his scheme for constructing docks at Calcutta. He requested the Company not to approve Mr. Tolly's plans. See Mr. Tolly's Petition, IOR/D/149/ff 65–66: 1773–82. Gokul Ghoshal Bhukailash also brought a litigation against him. See Chakraborty and De, *Select Documents on Calcutta*, xvi.

[15] Wet Docks IOR V/26/750/2, 8.

the members planned a new survey, which was conducted over six months by the private company Messers. Burn and Co., working alongside the Surveyor General's office. They pegged the value of the entire area stretching from Tolly's Nullah to the banks of the river in Khidderpore at 1,366,600 rupees. According to their estimates, since the Company would need only around 498 bighas of solid ground, the cost of land acquisition was calculated as no more than 593,000 rupees.[16] The British, as we learned in the preceding chapter, did not always compensate for acquiring urban swamps. The Committee argued that they would quickly recoup this sum through taxes.

While profit motives were moot in the wet dock construction project, a new rubric began organizing the infrastructural activities of the Company. New economic imperatives began to yoke people and places together under a different organizing principle. The joint-stock company's infrastructural ventures with stakes in both the risks and profits replaced private speculative projects of the earlier decades. While storms initiated a discussion around riverine safety, it was an emerging field of economic principles, which had begun reorganizing rural and urban lives, that bolstered the idea of constructing wet docks in Calcutta. Robert Torrens's ideas of organizing spaces along principles of imperial *Zollverein* (zones of taxation) – which meant the creation of new economic zones without political federation – were already gaining ground throughout the colonies.[17] Plans to economize spaces through differing zones of taxation impacted Calcutta's land market. These plans generated new monetary valuation through novel calculative registers, thereby extracting urban properties from earlier systems of values: ancestral, sacred or social. While Lacam's, Watson's and Colvin's projects of harbor and dock-building sought to funnel their excess capital from trade into future capital accumulation, the joint-stock company's projects created new spaces for the reproduction of British capital. Instead of waiting on profits from individual public works, the colonial officials could order spaces within the city under single tolls, taxes and dues.

Within this new spatial conglomeration, draining and the management of water within the city would reconfigure the urban land market through self-regulating principles of the colonial economy. This is precisely why

16 Wet Docks IOR V/26/750/2, 13.

17 Lionel Robbins, *Robert Torrens and the Evolution of Classical Economics* (London: Macmillan, 1958), 225–31; for an application of Torrens's ideas in India, see Goswami, *Producing India*, 44.

the Secretary raised the question of whether the joint-stock company could have vested (commercial) motives for acquiring and transforming land in order to pocket the profits from it. That is also why Mackenzie, who shepherded the Strand Bank project, had to loudly proclaim that pecuniary interest did not drive that project. Through infrastructures and new forms of taxation, the colonial state sought to transform the existing order according to a set of principles that would mask the separation of the economic from the political and social life of the city's infrastructure.[18]

The hydraulic infrastructure of Calcutta created a market of its own. It is interesting that some of the major studies focusing on infrastructure and public works in the colony have primarily studied them from the perspective of colonial ideology and architectural history without subjecting them to the scrutiny of the markets and economies they created.[19] Once we begin subjecting our analysis of the hydrological remapping of the city to understand how they were linked to the larger economic transformations of the time, a different history emerges, one that moves beyond epidemiology as an organizing principle. For instance, in the case of wet docks, the members of the Committee applied Scottish economist John Ramsay McCulloch's novel theories of commercial principles to Calcutta.

Trained in the Ricardian School, McCulloch's ideas were already making rounds within the Company circles in the 1840s. His works about commerce impacted not just the construction of the quays in London, but also the wet docks in Calcutta. McCulloch's *A Dictionary, Practical, Theoretical and Historical of Commerce and Commercial Navigation* (1832) argued that, prior to the construction of wet docks on the banks of the Thames, property annually "pillaged from ships in the river, was estimated at £500,000 per annum."[20] While he mentioned that the number could not be fully trusted and was perhaps inflated, it demonstrated the loss incurred on riverine trade due to the absence of a dock in

18 Karl Polyani argued that masking the separation between the economic aspects from social and political lives was central to the production of the market from the late nineteenth century. Karl Polanyi, *The Great Transformation: The Political and Economic Origins of Our Time* (Boston, MA: Beacon Press, 1957), 68–69; for an adaptation of the Polyani's argument in colonial India, see Sen, *Empire of Free Trade*, and Birla, *Stages of Capital*.

19 For an exception, see Ravi Ahuja, *Pathways of Empire: Circulation, "Public Works," and Social Space in Colonial Orissa (c. 1780–1914)* (Hyderabad: Orient BlackSwan, 2009), 224–58. He attempts to locate the development and commercialization of communication and canals in Orissa within the framework of a Lefebvrian "social space."

20 Wet Docks IOR V/26/750/2, 20.

London. The Mercantile Members of Calcutta, borrowing his ideas, pointed out that after the partial opening of West India Docks in London's Thames in 1802, the annual pilferage disappeared, which meant there was an annual profit of 1¼ percent. This curious percentile logic was brought to bear on the proposal for Calcutta's wet docks. Melding the economic logic with projects of spatial rearrangement was becoming more salient than narrow profit calculation. Through the consolidation of this logic within the bureaucracy of planning the mobility of the landscape and the very memory of a soaking space was gradually erased.

McCulloch's account of pilferage was hastily stretched across the Indian Ocean and applied to the Hooghly River. The shape-shifting character of numbers in the hands of these early colonial officials becomes clear when we see that the ratio of pilferage to profit in London's waters was abstracted into a percentage with universal tendencies that, if it held true for London, must be true for Calcutta. Mercantile Members built the percentage into their budget calculations, not as pilferage, but rather they converted it into expected revenue since wet docks would eliminate existing or purported theft from the waters of the Hooghly. Hence, following the report: "If then it was even supposed that such a loss as 1¼ per cent was sustained from pilfering on the Thames, where there is a well-organized river police, it seems safe to assume 1 per cent for the calculation of the amount of pilfering in the Hooghly, where the very men employed to protect property are found to plunder it, in those few instances where detection takes place."[21] People living and working in the swamps, like Lacam, could not be trusted.

Numbers and facts generated in England were now applied to the colony to justify the profitability of infrastructural projects. In the process we also begin to witness a movement away from the mercantile logic in urban spaces to a new episteme that began to organize colonial spaces as sites for the reproduction of British capital, which were part of and an extension of the imperial space.[22] This new episteme, according to scholars, was influenced by the growing ascendency of Utilitarian principles following the end of the East India Company's monopoly in 1813 and the formation of political economy as both a scholarly and a colonial bureaucratic discourse.[23]

[21] Wet Docks IOR V/26/750/2, 20.

[22] Goswami, *Producing India*, 44. See also Giovanni Arrighi, *The Long Twentieth Century: Money, Power, and the Origins of Our Times* (London: Verso, 2010).

[23] Goswami, *Producing India*, 43–44.

As this particular economic logic was replacing earlier mercantile profit-driven ventures, such logic was also standardizing and eclipsing long-existing practices along the riverbanks. The Wet Docks Committee proposed that, if inland trade moved away from rivers to the proposed railways, and maritime trade concentrated on the newly constructed wet docks, then the Company would successfully restrict "pilfering of cargoes which so constantly arises, and especially in this port, from the employment of boats, besides the great loss of time they occasion in a river with strong spring tides."[24] What appeared as pilferage and delay within this emerging Ricardian logic of profits might have held other meanings and existed within other structures of law. Prior to the arrival of the British, tolls and taxes on waters in Bengal were managed under different legal registers. According to the Hanafi canon laws that the Mughal Empire practiced in Bengal and elsewhere, imposing taxes and duties on water was prohibited. Mughal laws also stipulated that canals, rivers and other water bodies had to be cleared and repaired at the expense of the public treasury.[25]

It would be wrong to assume that riverine or maritime trade was not a source of revenue in the Mughal and early modern periods. From the seventeenth century on Mughal Emperor Akbar established a department of *mir-i-bahr* whose duty it was to police riverine trade and the loading and unloading of goods, levying export and import duties and managing Haj traffic.[26] With this move Akbar consolidated the earlier *nawwara* (river fleet) system, which had been in existence in Bengal from the thirteenth century.[27] In Early Modern Bengal and the Indo-Gangetic plains the office of the *mir-i-bahr* (or *chauki* as it was sometimes known) at the ports (*bandars*) was often farmed out to people who had won the favor of the Nawab of Bengal and they had the right to the duties on specific commodities of trade. Some of the duties levied in Bakshbandar, a port on the Hooghly, included the port or ferry duties on export and import (*mir bahari*), customary dues from boats (*hastnag*), a levy on building new boats (*nau sakht*), duties on boat-building material and all written passages or permits (*parvanas* and *dastaks*). Given Bengal's expansive network of riverine trade there were many intermediary positions between the

[24] Wet Docks IOR V/26/750/2, 12.

[25] Naqvi, *Urbanisation and Urban Centres under the Great Mughals 1556–1707*, 18–19.

[26] Naqvi, *Urbanisation and Urban Centres under the Great Mughals 1556–1707*, 19–20.

[27] Khondkar M. Karim, "Mughal *Nawaara* in Bengal," *Journal of Asiatic Society of Bengal* 14, no. 1 (1969): 51–59.

merchant and the Customs Department. For example, the leader of the boat (*ghat majhi*) wielded considerable power in securing duties (*ghat bhara*) from ferry customers and travelers for the movement of goods and for securing boats and ferries for merchants, traders and landlords.[28]

The entire world of customary dues and exchanges appeared to the colonial officials as nothing but "pilferage." Removing pilferage and streamlining taxes would not only remove a whole range of intermediaries who lived off the water and the riverbanks along Calcutta, but it would also bracket the river into a contractual resource that was contrary to existing practices in the Gangetic waters.

Murray Gladstone, a merchant and one of the Mercantile Members, proposed two new forms of revenue from the docks, which he argued should be used to defray the costs of construction. First, he proposed that duties be levied on shipping and all vessels, with the exception of *dhonie* (fishing boats) and coasting crafts. Such duties would be fixed by the Company according to the length of the voyages and frequency of presence in the Calcutta port. Second, he wanted to introduce "wharfage rates," which would be a form of duty imposed on imports into and exports out of Calcutta.[29] This would bring a considerable amount of money to the Company. During this same period, Britain was exporting seven staples for which Calcutta was serving as an emporium. These staples were indigo, sugar, saltpeter, rice, flour, jute and linseed.[30] The Company's estimated profits through these charges based on the riverine and maritime traffic would be approximately 615,490 rupees, which was a significant amount.[31] However, this was just the profit to be had from revenue. To this the Mercantile Members added the estimated pilferage, which, based on the London percentage, came to 1,029,472 rupees and made the total profits as large as 1,644,962 rupees. The Wet Docks committee admitted that the amount of estimated pilferage might be "enormous, and may possibly be considerably in excess of the reality, but there can be no doubt that the actual extent of depredation is very great."[32]

Once the economic logic was aligned with the political will for the wet docks in Calcutta, the final obstruction that arose was how to build on

[28] Sen, *Empire of Free Trade*, 38–42.
[29] Wet Docks IOR V/26/750/2, 15.
[30] Wet Docks IOR V/26/750/2, 16.
[31] Wet Docks IOR V/26/750/2, 17.
[32] Wet Docks IOR V/26/750/2, 20.

"bottomless alluvion."[33] Now that these indeterminate ecological spaces had been geographically classified as alluvion and legally zoned as accretions, managing them became merely a matter of scientific and engineering considerations. As engineers of the Bengal Army began to construct the docks, they had to fight the nature of the subsoils in the city, and solve the problem of how to keep the roadway connecting the docks to the city from turning into a swamp.[34] However, unlike the Company's previous adventures with alluvion, this did not unravel its bureaucratic system.

Though difficult to solidify for construction purposes, by the middle decades of the nineteenth century alluvion became merely an engineering concern. This was the powerful impact of colonial law on the environment. Army engineer W. N. Forbes was initially baffled by the fact that, in spite of his long experience of building on bottomless alluvium, he had still not been able to work out a system to keep the roads of Bengal from turning into drains during the monsoon, or for that matter to keep drains from becoming dry patches during winter. Writing to F. W. Simms, surveyor of Calcutta, in 1846, he remarks that

> it requires to be noticed, that in Bengal no description of road can be kept dry without having alongside of it, drains or excavations of such sectional areas as to afford the material for embanking it, and it may also be observed that these excavations are usually either so deep as to become in the rains water courses speedily ruinous to the road, or *so broad* as to form long shallow lakes – necessarily occupying (at great first cost for ground) a large surface of land.[35]

Placing railways over the Tolly Nullah to connect the markets to the dock would require working at an elevation of fifteen feet, keeping the "season of Inundation" in mind.[36] While the British had become used to seasonal roads, railway technology on the mobile landscape posed its own problems.[37] Elevated archways for the rail lines were proposed as a cheaper solution compared with the costs involved in draining and keeping the

[33] "Letter from Lieutenant Colonel Forbes to F. W. Simms," 16 July 1846, Supplement to the Report on Wet Docks IOR V/26/750/2, 7.

[34] D. D. Cunningham, "Note Regarding Certain Characters in the Sub-Soil of Calcutta," in *Scientific Memoirs by the Medical Officers of the Army of India*, Part III, ed. Benjamin Simpson (Calcutta: Superintendent of Government Printing, 1887), 17–25, IOR/V/25/850/3.

[35] "Letter from Lieutenant Colonel Forbes to F. W. Simms," 16 July 1846, Supplement to the Report on Wet Docks IOR V/26/750/2, 6.

[36] "Letter from Lieutenant Colonel Forbes to F. W. Simms," 16 July 1846, Supplement to the Report on Wet Docks IOR V/26/750/2, 3.

[37] Iqbal, *The Bengal Delta*, and Ritika Prasad, *Tracks of Change Railways and Everyday Life in Colonial India* (Cambridge: Cambridge University Press, 2015), mention how the

tracks dry throughout the year. Forbes proposed that the elevated archways be fortified and constructed with iron-tied masses of masonry that would be made of "well-burnt brick-earth such as, throughout India, is almost everywhere met with."[38]

However, the granite or iron-like brick, which was purportedly so ubiquitous in India, did not solve the problem of swamping that threatened to undermine the entire public works. D. D. Cunningham, a member of the Sanitary Commission in Bengal writing in his scientific memoir almost forty years after Forbes's declaration about cast-iron bricks, documented how the dock construction unraveled experiments on the subsoil of Calcutta. The subsoil of Calcutta was something that had continued to baffle geographers, surveyors, members of the sanitary department and the Bengal Army engineers. Cunningham writes:

> When the question of the construction of the docks was still under consideration, it was authoritatively announced that the subsoil in which it was intended to lay the foundation consisted of "blue clay," that it contained no subsoil water, and that in its impermeability almost equaled that of cast-iron, so that no accumulation of water would ever occur in the excavations necessary for the construction of the dock beyond that due to direct rainfall so long as means were taken to exclude surface drainage.[39]

Reaching a stage of impermeability in Calcutta's subsoil had to be achieved by exposing its moist character to sunshine and evaporation. In spite of the experience of working primarily with the swamps and peat bogs of Great Britain, the engineers knew very well that Calcutta had a surplus of sunlight that would help attain that state. Timing the excavation would be the trick to solving the problem of water, not the purportedly ubiquitous blue clay. Cunningham proclaimed that deep excavations must take place in summer, or else the municipal department would incur huge expenses.[40] Through multiple experiments, the British finally constructed a massive system of wet docks in the southwestern part of Calcutta, from which the city officials continued to reap enormous profits, all the while raising the value of lands around it.

railway construction interrupted the riverine flows in the Gangetic plains resulting in devastating ecological consequences.

38 "Letter from Lieutenant Colonel Forbes to F. W. Simms," 16 July 1846, Supplement to the Report on Wet Docks IOR V/26/750/2, 7.

39 Cunningham, "Note Regarding Certain Characters in the Sub-Soil of Calcutta," IOR/V/25/850/3, 17.

40 Cunningham, "Note Regarding Certain Characters in the Sub-Soil of Calcutta," IOR/V/25/850/3, 24–25.

In analyzing the management of urban ecology of Calcutta we witness that infrastructure became the empire with the legal regime as its justificatory window dressing. As we will see in the next section, the transfer of power from the East India Company to the colonial state was not so much a break, but rather it signaled the consolidation of "the shift, begun in the late 1830s, from mercantile to territorial colonialism."[41]

DRAINING AS URBANIZATION

Public works, which had begun to take a distinct shape in the middle decades of the nineteenth century, were constantly overwhelmed by the costs involved in working with the soaking ecologies of Calcutta. These public works projects became especially critical beginning in the middle decades of the nineteenth century, a period when colonial power attempted to make a claim of comprehensive territorial sovereignty.[42] How were such territorial consolidations achieved? According to Goswami, these territorial projects "involved attempted monopolization of regulatory powers by an increasingly centralized apparatus, the development of an elaborate, hierarchical bureaucracy that surveyed, mapped, and measured both land and people, the deepening and widening of the administrative and military reach of the state, and a determined reinvestment in epistemic modalities of rule."[43] While surveying, mapping and measuring began more than a century prior, it was scaled up in the production of a more bounded homogeneity to create what she called the "colonial-state-space." Grafted upon this emerging colonial-state-space, the taxations on urban spaces and properties laid the groundwork for the emergence of a speculative real estate market in Calcutta, and also in Bombay, Rangoon and other cities of the Raj.

The speculative urban land market was birthed at the crucible of legal and economic spatial practices of the colonial state. After the Crown took over, the British Raj restructured the fiscal aspects of the urban landscape, intensively creating various pockets of revenue generation and a form of private wealth accumulation. This restructuring took place at a time when

41 Goswami, *Producing India*, 43.

42 Yet if one were to follow the idea that the corridors and enclaves of legal cartography posed limits on colonial state's power, as repeatedly argued by Lauren Benton and others, one would perhaps be left with a more patched territorial sovereign space than the bounded colonial-state-space that Goswami argues for. Benton, *A Search for Sovereignty*, 2–11.

43 Goswami, *Producing India*, 31.

the indigenous populations, as seen through the examples of Seal and Deb Bahadur, were bitterly and noisily staking their claims, thereby producing ethico-political subjective positions within the emergent nationalist frames. The earlier projects, such as Lacam's failed harbor construction and the Territories Department's successful execution of the Strand Road, exemplified the workings of the quasi-state body of the Company. They showed how the merchants and officials of the Company managed the landscape through technology and law. However, if these projects were openly implicated in profit-making ventures, where the profits were remitted back to the Company's purse, how did those very same operations emerge differently in the period after the Crown takeover?

The processes through which the nineteenth-century urban space was manufactured therefore offers us a canvas for recovering how this period was marked by maintaining and institutionalizing a productive disjuncture between the political and economic structures that produced imperial spaces as part of and yet differently connected to Great Britain as a global economic power.[44] This difference was critical in applying metropolitan economic logic to colonial spaces, yet subjecting the very same spaces to a different juridical logic. Above all this logic shaped the markets that connected the metropolitan centers to the colony, while simultaneously sustaining the "rule of colonial difference."[45] Analyzing the physical and social infrastructures and their creation in this backdrop offers new ways of understanding the urban space without reducing it to a site of biopolitical governmentality.

Let us then turn to the unique political economy of infrastructures to understand the market in which they were embedded. Investment in infrastructure across the empire witnessed a staggering increase following the Crown takeover in 1857. The East India Company had invested only £1.43 million of its average annual revenue of £20 million, an estimated 1 percent of its revenue from 1830 to 1857.[46] After 1858, state debt was channeled toward massive investments in both social and military infrastructure. Infrastructure here included extensive railways, bridges, canals, irrigation, ports and postal services, all of which required the establishment of new institutions for producing, ordering and maintaining these

[44] Goswami also pointed out that while the production of the nationalist and colonial territoriality of the Indian space must be studied with the globalizing tendencies of British economy, it was not a unitary production but a multiplex of factors "creating contradictory force-fields." *Producing India*, 33.

[45] Chatterjee, *The Nation and Its Fragments*.

[46] Based on Goswami's estimates; see Goswami, *Producing India*, 45.

massive networks of transport and communication.[47] The research into canals and railways by Ravi Ahuja and Ritika Prasad shows that these infrastructures bounded the colonial territory while simultaneously differentiating spaces physically and internally through the creation of hierarchies of access and use.[48] Railway and canal infrastructures functioned in creating particular centers of production, distribution and consumption by connecting certain markets while bypassing others.

What can the study of urban infrastructures during this period add to this understanding of spatiality? If railways, telegraphs and canals restructured the movement of labor and commodities, then urban infrastructures created markets in land and properties in water, forging and sustaining the "territorialist and capitalist logics of power."[49] This expansion and the development of new modes of creating value and zones of capital accumulation within the cities also created new systems of credit and new modes of investing within the urban space.[50]

For the docks and the city to function without seasonal vagaries, another major draining project was proposed – this was the project to drain the Maidan, the open green space in the heart of the city, and to create a circular bund, or embankment, to channel the water. This would neatly divide the swamps surrounding the Maidan from the built environment of Calcutta. Investigating the Maidan drainage scheme helps us delve deeper into the final decades of the nineteenth century, when drying the land shifted from management of water as a means of protecting British investments to a site of revenue generation.

This was not going to be a matter of mere drainage and epidemiology, but had to be a project that transformed swamps literally into speculative properties. In order to do this, the Company sought the help of engineers from another swampy city with a speculative land market: Amsterdam. Around the same time, when Forbes was communicating with Simms about building on bottomless alluvion, the Chief Magistrate of Calcutta was planning to send two or three young engineers, trained at the newly established Addiscombe College of Engineering in England, to Amsterdam to study how to deal with encroachment of the sea

47 For the development of the railway and telegraph infrastructure of the empire, see Ian J. Kerr, *Building the Railways of the Raj: 1850–1900* (Delhi: Oxford University Press, 1995). Prasad, *Tracks of Change*; Deep Kanta Lahiri Choudhury, *Telegraphic Imperialism: Crisis and Panic in the Indian Empire, c. 1830* (Basingstoke: Palgrave Macmillan, 2010); Ahuja, *Pathways of Empire*.

48 Prasad, *Tracks of Change*; Ahuja, *Pathways of Empire*.

49 Arrighi, *The Long Twentieth Century*, 58.

50 Chapter 5 will elaborate on this.

into the city before they could be employed to work in Calcutta.[51] Amsterdam was not simply famous for managing water, but also, it must be stressed here, had one of the most lucrative land markets in nineteenth-century continental Europe.[52] Following this proposal in 1847 the Chief Magistrate of Calcutta laid out a comprehensive plan to drain the central parts of Calcutta.[53] These engineers, it was anticipated, would return with a specialization in turning swamps into speculative property. From the existing records it is hard to assess whether the engineers were successful, but Calcutta's land market was set to become a profitable investment soon enough. Epidemiological concerns and economic logic deemed the draining project expedient and necessary for the functioning of the city. Thus, the Chief Magistrate of the city began massive projects and investments into canalizing and piping water, projects that in the past had often been piecemeal in nature.[54] The garrison engineers of Calcutta (the Maidan being a fortification under army jurisdiction) proposed six culverts, or drain bridges, to turn the marshes into a "green space."

The imagination of what a "natural" urban park would be in a city had a distinct character in Calcutta. The open green space of the Maidan was "land" that belonged to the Army. However, it turned into a land–water admixture or a sheet of water during the monsoon months. This phenomenon of seasonal soaking that the monsoons subjected the urban landscape to was an aberration to the British residents in India. If the disappearance of *nala* (seasonal freshes, or spill channels) in the agrarian world was a cause of concern, a similar consternation marked the management of the Maidan. In the initial years, this moat-like quality of the Maidan formed the glacis to the Fort and served the purposes of fortification. However, the Maidan was also becoming a thoroughfare for the growing population of the city.[55] Calcutta, therefore, needed uninterrupted green between the city and the fort, instead of the swamp-like

[51] Judicial [Judicial], 9 January 1843, no. 8, WBSA, Kolkata.

[52] Clé Lesgar, *The Rise of the Amsterdam Market and Information Exchange: Merchants, Commercial Expansion and Change in the Spatial Economy of the Low Countries, c. 1550–1630* (New York: Ashgate Publishing, 2016).

[53] Judicial [Criminal], 29 December 1847, no. 26, WBSA, Kolkata.

[54] For instance, the canal Tolly Nulla was constructed initially as Tolly's private property. I will not be engaging with the epidemiological history of the city, nor with the planning history of piping and sanitation. For more on that, see Pande, *Medicine, Race and Liberalism*, and Dutta, *Planning the City*.

[55] Bhattacharyya, "Fictions of Possession," 142–66.

vegetation it had at that time. Moreover, as Chattopadhyay mentions, attempts to create an urban vista in the colonial city of Calcutta were orchestrated on the Maidan through a particular picturesque genre of colonial paintings that fueled nineteenth-century planning ideologies. The world imagined in those paintings then needed to be translated into reality through draining.[56]

Given the critical symbolic and economic position of the Maidan within the mid-nineteenth-century imperial geography, the Bengal Army engineers who established themselves as experts in constructing on the clayey silt of Bengal began to devise an intricate plan to drain the Maidan. According to their plan, the Maidan was first divided into distinct sections, each draining to one outlet. New and straightened draining outlets were constructed as they were deemed to "afford the readiest escape for the waters carrying the drains past the tanks instead of through them."[57] Rationalizing the draining system bypassed the natural drainage networks existing within the city. As I have mentioned elsewhere, man-made canals were not a British import, but during the precolonial times movements of fish were followed and recorded in songs and anecdotes to understand and build upon the land and waterscape of this space. Fishermen dug canals across this region following the seasonal movements of the fish. Those reticular movements and patterns of human labor on the space are now etched in the names of places as the only "archival" records from that period.[58] That particular way of knowing and apprehending geography began to disappear under the onslaught of the new knowledge of colonial regimes in which understanding and modifying the landscape was the purview of the Army engineers.[59]

Avoiding the excessive expense that came with producing a dry Maidan obligated the engineers to devise an intricate system of sluice gates, on the one hand, and to manage cattle grazing practices, on the other, all within the auspices of one plan. The biggest threat to the proper functioning of the drains in and around the Maidan was not just the annual rains, but also the cattle traffic over the drains.[60] The plan

[56] Chattopadhyay, *Representing Calcutta*, 43–49; see also Bhattacharyya, "Fictions of Possession," 142–88.

[57] Judicial [Judicial], 12 December 1864, no. 3812: 107, WBSA, Kolkata.

[58] Bhattacharyya, "Geography's Myth: The Many Origins of Calcutta," 151–53.

[59] Most early surveyors and hydraulic engineers were attached to the Army.

[60] Mr. Rayner, a garrison engineer, was asked to draw up a report of the existing drains on the Maidan, and he bitterly complained about the destructive force of the cattle

prescribed new rules regarding the movement of cattle. Mid-nineteenth-century paintings of the Maidan had already emptied it of animals, especially cattle that would graze there, to render the pictorial representation commensurate with the genres of picturesque that were developed in the English countryside.[61] With the expansive draining ventures, the Chief Magistrate laid down new rules to control cattle grazing on the Maidan in "order to prevent the drains being trodden into deep holes and becoming a series of irregular pools."[62] Subsequently, cattle grazing on the Maidan were taxed and turned into a source of revenue as the new classificatory spatial regime of the urban space began to take shape under the Raj.[63]

The annual monsoons also wreaked havoc in Calcutta by thwarting any attempts at transfiguring the marshes into any semblance of urbanity. The piecemeal drains of the earlier generation had to be replaced by scientifically planned drains. Lieutenant Colonel W. D. Short, the Chief Engineer of Bengal, thus devised his drainage plans by calculating the maximum threshold of rainfall to be "12 inches in 24 hours."[64] Such estimation and approximation was increasingly becoming a strategy for dealing with the messiness and instability of the fluid and seasonal geography of the city. Following that calculation of rain, the Maidan was divided into three portions, and the possible amount of water that might collect in each portion was calculated down to the inch. The drained water must go somewhere. Forbes, along with his fellow engineer Short, thus proposed an intricate array of canals and tunnels to rid Calcutta of all forms of the so-called moist and pestilential elements, as the mid-eighteenth century municipal language so overwhelmingly refers to it.[65] After multiple plans and surveys, they finally proposed that the draining project proceed along two lines: one would be the excavation of canals, and the other would be the leveling of the Maidan, especially where it

movement on the loamy and soft alluvium of the various open spaces in the city. Judicial [Judicial], 12 December 1864, no. 3812: 108, WBSA, Kolkata.

61 For the application of the English picturesque to the colony, see Guha-Thakurta, "The Compulsions of Visual Representation in Colonial India,"108–39; Pinney, *Camera Indica*.

62 Judicial [Judicial], 12 December 1864, no. 3812: 107, WBSA, Kolkata.

63 Bhattacharyya, "Fictions of Possession," 172.

64 Judicial [Judicial], 12 December 1864, no. 3812, WBSA, Kolkata.

65 Judicial [Criminal], 29 December 1847, no. 26, WBSA, Kolkata.

"lacked a natural slope" so as to turn the rainwater into spill channels flowing to the newly excavated canals.[66]

The project of producing "nature" and "green space" within Calcutta proceeded at the interface of engineering and hydraulic science. Yet, such a science was limited in the numerical rubric it employed to understand how the hydrological landscape functioned. The average calculation of rainfall was nothing more than a fictitious average derived from rainfall calculation models based on temperate climates in England. A similar inability to calculate average rainfall in the way one could for England created a problem for the physiocratic agrarian land settlement in 1793.[67] Moreover, just as Lacam's planning of his harbor was based on a selective geographical perception of land-water-relations, one could also argue that a similar blinkered vision dogged the creation of the Maidan as a green park. Instead of the soaking ecology of marshes with seasonal water channels that defined this topography, the city officials sought to produce a dry urban park. They assumed that spaces were atemporal, and such landscapes of flux had no place within the colonial urban space. Assumptions apart, even today the Maidan continues to become a sheet of water during monsoons.

THE SWAMP AND THE CHANGING ECONOMY

Where and how does this history of draining the swamp that was Calcutta fit within the emergent frames of the colonial-global economic regimes during the middle decades of the nineteenth century? What can the operation of these infrastructural projects teach us about how changes in colonial economy and law impacted the urban space and its land markets?

The space of Calcutta was zoned into various pockets of taxation following the 1857 Crown takeover of the colony – a period mired in economic crisis. As numerous scholars have noted, new legislation was introduced in the aftermath of 1857.[68] Some of this legislation critically reshaped the urban land market, especially the new Act for Imposing Duties on Profits Arising from Property, Profession, Trades, Office, Act

[66] Judicial [Judicial], 12 December 1864, no. 3812: 108–9, WBSA, Kolkata.
[67] Hill, *River of Sorrow*, 132.
[68] Radhika Singha, *A Despotism of Law: Crime and Criminal Justice in Colonial India* (Delhi: Oxford University Press, 2000); David Washbrook, "Law, State and Agrarian Society in Colonial India."

XXXII of 1860. This act transformed urban property into a taxable entity.[69] The immediate backdrop for this act was the ill-executed public-works ventures in 1850 and the famines through the 1870s, which depleted state revenue. Within a decade of passing the 1860 property tax, the British government passed the India Income Tax Act of 1886. This was in response to the depreciation of silver in the early 1870s, subsequently worsened by the bloated military expenses due to the Second Afghan Wars of 1878–80.[70] To tackle the crisis, the British parliament unleashed a whole host of economic laws on the Raj. The 1886 Income Tax Act for the first time consolidated the nonagricultural sector within a new fiscal arrangement. The towns, cities and their resources, namely the businesses and properties, had to generate revenue for the colonial state.[71] Ritu Birla described the late nineteenth-century fiscal reconstitution of the colonial state as a moment when "the foundational statutes of the 1880s ... confirmed a century-long shift from a physiocratic model of political economy focused on land as a source of revenue and wealth, to the taxation of income and investments."[72] Land does not disappear from the equation, but as I show in the final chapter, it enters the debate in a new manner completely shorn of its fluid history.

During this period, the British Empire began following a two-pronged fiscal policy. One was to either lower or abolish customs duties to maximize free trade within the empire, which resulted in the emergence of a new moneyed population. The other was the expansion of the fiscal system beyond agricultural or productive land to urban centers. As capital accumulation in select pockets of cities like Calcutta and Bombay was increasing, so was the question of expense becoming a centerpiece for the plans for drainage and canalization. Thus we see that, even as the members of the City Magistrate reached consensus for draining the city, after much budgetary deliberation it was found that these infrastructural projects would leave the Conservancy Committee, in charge of carrying

69 "Act for Imposing Duties on Profits Arising from Property, Profession, Trades, Office, Act XXXII of 1860," 19th Century House of Commons Sessional Papers, vol. XLIII.201, 7 March 1861 (British Library, London).

70 John Dacosta, *The Financial Situation in India, as Exposed in the Budget Statement for 1886–87* (London: P. S. King and Son, Parliamentary Agency, 1887), India Office Records, British Library, London.

71 Radhabinod Pal and Balai Lal Pal, *The Law of Income Tax in British India: Being Act XI of 1922 as Amended by Act VII of 1939. With Explanatory Notes and Commentaries* (Calcutta: Eastern Law House, 1940).

72 Birla, *Stages of Capital*, 36.

out the task, with a decade-long debt of 2,052,627 rupees.[73] Raising money through the lottery under the aegis of a Lottery Committee, as was the early nineteenth-century trend in the colonies and the metropole, had been outlawed since 1840s, because it encouraged the urban population to indulge in gambling.[74] One would imagine that the Calcutta Municipal Corporation established in 1876 would have at its disposal a budget for funding infrastructural projects instead of either relying on lottery or coercing private individuals to fund its projects. This was unfortunately not the case. Money for public works was slowly becoming scarce and earlier forms of financing urban infrastructure were becoming rarer simultaneously. The municipal officials were constantly forced to secure new sources of taxation in properties, assets and professions.

Alongside earlier methods of coercion and extralegal land acquisitions through which the Company and later Crown constructed the Strand Road, the Calcutta Municipal Corporation now needed to devise new spatial practices that would bracket space and institutionalize revenue generation. The emergence of a new-moneyed population therefore offered an opening for systematizing fund extraction beyond voluntary and forced philanthropic ventures. It is in these moves that we see a changed state structure of corporate governance begin to emerge, one that would eventually consolidate the soaking ecology into bounded dry land, where water would be contained in canals, tanks and drains.

As early as 1847 the then Secretary of the Government of Bengal F. J. Halliday had written to the government of India that plans had to be devised to get the public to raise money, "if not the whole, a considerable part of sum," that would be necessary to buy land for various town planning projects in the city.[75] He was confident that money could be raised through subscriptions from the Calcutta public, as they had been forthcoming with their donations for the construction of the hospitals, which added an unanticipated amount to the working budgets of the medical establishments. Halliday, speaking from experience of what the Calcutta public would make available, reflected on the possibilities of changing that ad hoc nature of endowment, gifts and subscriptions into a steady flow of revenue to the municipal coffers.

Halliday proposed a range of sources to fund the massive infrastructural projects. The city already had a number of sources of revenue from

[73] Judicial [Criminal], 29 December 1847, no. 26, WBSA, Kolkata.
[74] Datta, *Planning the City*, 28.
[75] Judicial [Criminal], 29 December 1847, Prog. 26, WBSA, Kolkata.

the following items: annual house taxes, *akbaree* (tolls from markets), tolls on the Calcutta canals, ground rent, fees, fines, escheats, rent from the Chowringhee Plains and the products made by convict labor. Halliday wanted to augment this with two other sources of income: additional taxation on houses and a carriage tax. Thus we see that, much before the Act of 1860, which transformed urban properties into taxable entities across the empire, the Company already had begun to raise a considerable amount of money through tolls and taxes on urban property. According to Halliday's calculation, the two new taxes would add a surplus of 200,000 rupees for the city officials.

Halliday also proposed new bureaucratic positions of urban governance whose task it would be to administer the taxation and oversee the new relations developing in the city between private capital and its public works. Creating new officiating positions was an initial step toward institutionalizing certain channels of money circulating within the burgeoning land market. Moreover, according to Halliday's plans, these new positions would be integrated into a form of corporate governance. That meant creating an early form of corporation, which could then become a legal entity with rights to purchase and hold lands. This maneuver would remove the kinds of concerns that had dogged the early Wet Docks Committee on exactly what the role of a joint-stock company could be in these investments and how might the Company claim the profits arising from them. Hence provisional bodies like Strand Bank Funds, which were given the power to hold the newly formed land on behalf of the Lottery Committee during the 1820s, were finally being institutionalized by the time of Crown takeover.[76] Halliday's foresight and recommendations would take more than half a century to take root in the city. These moves, along with the new financial instruments that the Crown unleashed on British India, restructured the economy of the city, creating pockets of speculative ownership.

"*Company'r latgiri, porer dhyone poddari*" (Company's bossiness is nosing in other people's property) – thus goes a famous Bengali doggerel verse from nineteenth-century Calcutta. It beautifully captures how meddling in other people's property and wealth constituted and consolidated the Company's political authority as a public agent of land in Calcutta.

[76] Judicial [Criminal], 29 December 1847, no. 26, WBSA, Kolkata. For a fuller account of the changing structures of municipal governance from Lottery Committees to the Justices of Peace, City Magistrate and the Calcutta Municipal Corporation, see Datta, *Planning the City*.

The making of Calcutta took place in the crucible of extraction, coercion and taxation on the Indian population. This chapter laid out these various shades of illegitimate and legitimate forms of accounting practices that were integral to the infrastructural and town planning projects. The political economy of lotteries and subscriptions, as well as voluntary and forced gifting practices, built the city. The relation between private capital and the construction of public goods therefore has a much longer genealogy stretching back to the days of Lacam.

Exploring the colonial urban setting through the regime of property and the land market illuminates the particular workings of what in contemporary times has been institutionalized as public-private partnerships in Indian cities. Yet we must not draw a direct line of causality from a private entrepreneur like Lacam to the "Calcutta Public," who, in Halliday's terms, were ever forthcoming with money. If Lacam wanted to insert entrepreneurial aspects within early modern forms of infrastructure in Bengal with hopes of monetary gain, by the mid-nineteenth century the story had been significantly transformed. From the latter half of the nineteenth century, the government in the form of the market mediated these circuits of private capital in the name of public goods, whose benefits and profits went to the colonial coffers. The emergence of new Indian actors would disturb the network of urban capital and the colonial land market during the early twentieth century. Such a disturbance in the land market would, therefore, demand a complete reorganization of municipal governance.

PART III

UNREAL ESTATE

5

Speculative Properties

From the turn of the twentieth century, a rhetoric of land's futurity in terms of value gradually replaced the watery past of the swamps of Calcutta. In the dried city, land became a precious commodity as the marshes were reconfigured into land-in-waiting for property development. This process of changing the valuation of land was not just specific to Calcutta but a global phenomenon, produced in the crucible of private landlordism, philanthrophism, regulatory projects of quasi-market bodies like Improvement Trusts, builders' lobbies and land syndicates, and the gathering momentum of housing rights and social welfare politics.[1] A fundamental reorganization of land value was already under way since the nineteenth-century infrastructural projects of draining and canalization that restructured the city into pockets of dried land and land waiting to be developed. In short, the ecological variability of Calcutta's landscape was becoming one increasingly marked by spatialities of (economic) difference. Such differences were written into the thriving land and speculative property markets, which fomented an urban housing crisis in the

[1] Builders' lobbies emerged in cities like Bombay and Calcutta as an institution to reckon with both politically and economically from the second decades of the twentieth century. This was not unique to the British Raj but a global phenomenon. Property owners' associations, similar to the builder's lobby, emerged as a formidable force in Paris during the same period. For the changes in the political and legal structures of ownership in Bombay, see Dossal, *Theatre of Conflict*, 216; Nikhil Rao, "Uncertain Ground: The 'Ownership Flat' and Urban Property in Twentieth Century Bombay," *South Asian History and Culture* 3, no. 1 (2012): 1–25; for Paris, see Alexis Yates, *Selling Paris: Property and Commercial Culture in the Fin-de-Siècle Capital* (Cambridge, MA: Harvard University Press, 2015), 59–98.

city, finally culminating in the Calcutta Rent Act of 1920 and the subsequent Calcutta Municipal Bill of 1925. Ecological difference did not disappear, but new visual literacies of viewing urban watery spaces and emerging economic possibilities erased the fluid realities of the swamp ecology.

Land, as a valuable commodity, charted myriad paths through different regimes of values and networks of circulation, morphing into new social, cultural and economic valences, thereby financializing existing forms of sociality. The materiality of value *in* land was slowly supplanted by the deferral of value to come *through* investments in land at the turn of the twentieth century. While the previous chapters focused on the process through which colonial law translated various spatial assemblages into property and how this newly-created landed property regulated social life, this chapter focuses on the economic circuits of land as property. Accordingly, the housing crisis of the twentieth century also highlights how urban housing entered a new politics of exchange and valuation, thereby generating novel politico-economic regimes and modes of "market governance."[2]

Focusing on the differences between property speculation and land profiteering, as perceived by municipal officials, brings out the immanent contradiction that marked the spatiality of colonial power. These contradictions can be summed up as the homogenizing spatial practices of colonial economy, which at the same time produced internally differentiated urban forms and material practices.[3] By the twentieth century, the question of property became subsumed into the rubric of housing, which became both a form and a framework for understanding the characteristics of urban land markets, the future valuation of swamps and the behavior of people who owned land or hoarded it. Thus, questions pertaining to controlling ecological risks of a littoral landscape were now superseded in the officialese as proper economic management of the financial risks in a speculative property market.

This was also a period when housing emerged as an urban form of global concern, which was also deeply hierarchical and fragmented, especially under the new instantiation within colonial power.[4] Property

[2] I borrow the term "market governance" from Birla, "Introduction," in *Stages of Capital*.

[3] Lefebvre, *The Production of Space*, 81; for a scalar reading of spatial politics, see Neil Brenner, "The Limits to Scale? Methodological Reflections on Scalar Structuration," *Progress in Human Geography* 25, no. 4 (2001): 525–48.

[4] Lefebvre, *The Production of Space*, 282; Daniel T. Rodgers, *Atlantic Crossings: Social Politics in a Progressive Age* (Cambridge, MA: Harvard University Press, 1998).

relations and the urban land market became organizing forces, zoning spaces, homogenizing ecology and simultaneously manufacturing people as agents in the economic assembling of these spaces. This began a process that was not merely a project of instituting a particular kind of (contractual) property relation in the colony, but rather involved the embedding of the property market within the larger forces of the imperial economy.[5] These propertizing tendencies begun over a century ago erased the hydrological memory of the city. Today these tendencies are manifesting themselves in our contemporary world as disappearing wetlands, shrinking coastlines of the estuary – spaces whose value is measured by the state in terms of real estate.

THE FUTURE VALUES OF LAND

On 3 September 1919, William Heath Phelps, the head of an industrial tailoring firm and a member of the Calcutta Municipal Corporation, complained that the cost of housing, a basic necessity, was increasingly subject to the laws of the market.[6] Objecting to the fact that this law had become "a fetish," Phelps criticized the commissioners for their professed inability "to see the difference in principle between the purchase of a pair of boots [considered an item of luxury] and the renting of a house."[7] The housing shortage was the fallout of a speculative land market. During the early decades of the twentieth century, the political mood of Calcutta, like other colonial cities in British India, was charged with two competing interests. On the one hand, there was a growing working-class movement for better working and living conditions, and, on the other hand, there were complaints from nationalists (who also happened to be landlords)

5 Anthony G. Hopkins, "Property Rights and Empire Building: Britain's Annexation of Lagos, 1861," *Journal of Economic History* 40, no. 4 (1980): 777–98.

6 "Corporation of Calcutta," *Amrita Bazar Patrika*, 6 December 1906, 5; Phelps objected to treating a nonessential commodity like boots as similar to housing, which was becoming a basic necessity. It was difficult to find a single commodity whose price had not gone up during this period across the entire spectrum of India. "Cloth, salt, kerosene, ghee, chilies and sugar all registered abnormally high prices in the middle months of 1918 and in some localities were simply unavailable. Cloth prices, two or three times their pre-war level." See for instance David Arnold, "Looting, Grain Riots and Government Policy in South India 1918," *Past & Present*, no. 84 (1979): 119.

7 Resolution Regarding the Appointment of a Committee to Enquire into the Causes of Excessive Land Values and High Rents in Calcutta, Municipal Miscellaneous Progs., April 1920, Kolkata: West Bengal State Archive (hereafter Rent Enquiry Committee, Municipal Miscellaneous Progs. April 1920, WBSA); "Calcutta Rents: Committee of Enquiry Appointed," *Amrita Bazar Patrika*, 16 October 1919, 8.

about what constitutes "legitimate need" and the limits of the land market. How did the housing crisis, fueled by rapidly increasing rents, produce a discourse where the swamps and wetlands of Calcutta emerged as land-in-waiting, and where building houses and propertizing swamps came to be accepted as a developmental project for the public welfare? How was this particular debate framed by the peculiarities of the emerging market in housing and land?

Almost a century and a half ago, the British were challenged by the fluidity of the aqueous territory. The denomination of alluvion, marsh, accretion, char, wetlands and swamps had gone through multiple iterations in legal, ecological and economic registers. By the time of the early twentieth century, the volatile nature of the land market meant that the British responded by devising ways of understanding the aqueous territories as housing-in-waiting. This chapter documents the process of this emergence.

Urban property and house rents were rising at an exponential rate across British India since the outbreak of the First World War. The towns worst hit by the housing crisis were Karachi, Bombay, Madras, Rangoon and Calcutta. In Calcutta alone there was a 38 percent rise in house rents in 1918 in the immediate aftermath of the war.[8] In the wake of rising prices and housing shortages, municipal officials were suddenly forced to designate urban housing and rent speculation as new objects of regulatory concern.[9] In 1919, under Phelps's chairmanship, a Rent Enquiry Commission was set up to inquire into the excessive rent and land values in Calcutta, which brought forth a series of debates in both the municipal and public domains.[10] Was housing to be treated as an item of choice – a luxury like a pair of boots – subject to rules of supply and demand, at a time when urban land and landed property were becoming critical assets of investment? Or was it to be viewed through the lens of rights and defined as a legitimate social need, like basic food and clothing? With the establishment of the Rent Commission, Calcutta's land and rental market gradually emerged as an object of knowledge and site of regulation for the governing officials within the municipal and judicial circles. Realizing the future value of swamps-as-housing became the organizing rubric for the municipal administration of land in the expanding city.

8 "War Prices and House Rents," 3.
9 "Calcutta Rents: Committee of Enquiry Appointed," *Amrita Bazar Patrika*, 16 October 1919, 8.
10 Rent Enquiry Committee, Municipal Miscellaneous Progs. §§34–45, April 1920, WBSA.

The abnormalities generated by the wartime economic restrictions following the World War I, coupled with grand visions of colonial town-planning partially implemented under the Calcutta Improvement Trust (CIT), established in 1911, gradually transformed the urban land and housing markets in colonial Calcutta.[11] Calcutta's housing market had been considered a lucrative business ever since the early nineteenth-century urban expansion. Rents were exorbitant, and even people in the middle-income group employed in the colonial offices considered landed property a worthwhile investment.[12] However, nothing about the earlier inflations in the urban land market precipitated a Rent Commission like the one of 1919. What precisely was unique about landed property and the rental market during this period?

In the postwar period the changes emerging within the colonial economy resulted in new modes of wealth accumulation as well as a new class of a landowners, who were increasingly castigated within the municipal debates as "extortionist landowners" and within the press as "house-bullies" in Calcutta.[13] The term "new landlords" in the official documents was often a euphemism for the Marwaris, or the North Indian commercial classes in the city.[14] Examining the official and popular debates that grew out of the peculiarity of the interwar period helps illuminate how concerns about the volatility of the liquid landscape were replaced by attempts to manage the mercurial land market. As a result, new ideas around investing and accumulating wealth in landed property began to emerge. At the same time, the so-called armistice regulations of the wartime economy were restructuring the existing landholding and social networks of power.[15]

[11] Bagchi, "Indian Economy and Society during World War One," *Social Scientist* 42, nos. 7–8 (2014): 5–27.

[12] Chattopadhyay, *Representing Calcutta*, 79–80.

[13] For the use of the term "extortionist landlords" and "extortionist rents," see Papers Relating to the Calcutta Rent Bill, 1920 IOR/L/PJ/6/763 (British Library, UK); "Calcutta Rent Commission," *Amrita Bazar Patrika*, 11 December 1919, 6.

[14] I am using the terms "indigenous" and "Marwari" to broadly speak about the various North Indian trading communities in order to reflect the archival language. The Marwari community was emerging as one of the predominant communities engaged in business in Calcutta around this time, and government documents and newspaper articles would use the terms "native traders" and "Marwaris," often interchangeably.

[15] The municipal debates use the terms "armistice" and "wartime regulations" or "restrictions" interchangeably. Rent Enquiry Committee, Municipal Miscellaneous Progs., April 1920, WBSA.

The colonial and indigenous reaction to the local housing crisis of Calcutta documents the consolidation of the various techniques of "market governance" that I touched on in the previous chapter. These techniques of governance operated by regulating the fictive and imaginative staging of land as an asset in the emerging housing market. As urban land increasingly became a marker of a stable asset of investment, a particular fiction of value accrued to land. Or, stated differently, land finally emerged as fictitious capital through the extraction of rent, as an entity for investment and through the imposition of housing and property taxes in the cities. The birth of landed property as a speculative commodity was intricately intertwined with projects of draining Calcutta, the ecological transformation of the city and the changes within its economic geography.

The municipal debates focused on the problem of land, which most European and some Indian landlords agreed had become the object of financial speculation and therefore entered a contested regulatory domain. New urban rent ceiling laws were proposed and debated across British India, especially in cities like Karachi, Calcutta, Rangoon, Madras and Bombay.[16] British municipal members and the Anglophone factions of Indian land developers proposed wartime regulations in the name of general public well-being and urban economic health. These regulations, born out of a rigorous grounding on principles of exchange, the members argued, would lay the foundation for sound financial decision-making by the "new landlords" who, according to them, showed a predilection for economic fatalism. Others, particularly the Bengali nationalist members who considered themselves the established landed gentry, contested the legal validity and financial legitimacy of such a proposal. Still other factions, primarily the British merchant class, including Phelps, argued that by calculating the cost of living and understanding the supply and demand of housing they could arrive at a market-driven property rent that could protect Calcutta's working classes, while at the same time securing increased tax revenues for the municipality from the growth in property prices. They proposed that numerical predictions about present

[16] Speculation was also rife in building and construction materials, laying the foundation for the emergence of the builder's lobby especially in Bombay; see esp. Dossal, *Theatre of Conflict, City of Hope*, 177–230; Nikhil Rao, *House, but No Garden: Apartment Living in Bombay's Suburbs, 1898–1964* (Minneapolis, MN: University of Minnesota Press, 2014). For specific costs in Calcutta, see section on financial report in the *Annual Report of the Calcutta Improvement Trust, 1920*; Richards, *Improvement and Town Planning of the City of Calcutta*, 20–25 and passim.

and future land values should be the basis for regulating profiteering in the housing and land markets in Calcutta.

The rental and housing markets emerged in Calcutta predicated on notions of housing rights as an outgrowth of worker protest, on the one hand, and land speculation, on the other. This tension opened up a space for colonial intervention, one that drew on numerical reasoning driven by discourses of financial calculations of value and potential risks within the emerging urban real estate market. The regularization of the urban land market and the rule of property worked in tandem to produce an "unlivable Calcutta." For some municipal members, namely those who spearheaded the founding of a rent commission in Calcutta as a response to the rising housing costs, the solution to this unlivability lay in municipal governance through economic management of land and housing. They proposed that this manner of management would also cure another evil, namely the cultural pathologies that had begun to hinder sound financial decision-making in the market, threatening the equilibrium of housing demand and supply.

RATIONALIZING THE LAND MARKET

The increasing unlivability of Calcutta also became a literary trope producing new aesthetics of urbanism. How much and how rapidly did the city change over the last few decades of the nineteenth century for new urban literary genres to emerge – both poetic and prose satires? The many aspects of urban life became a central theme for poet, writer and film director Premendra Mitra (1904–88). Affiliated to *Kallol*, a progressive writer's group founded in 1925, he wrote about the fickle movement of capitalist development in the property market that transformed Calcutta from the turn of the twentieth century in his 1934 novel *Agamikal* (*Tomorrow*). He narrates urban development through tropes of loss – the loss of ponds, water tanks and the water bodies of Calcutta. However, in Mitra's writing this was not only an ecological catastrophe in the making. What concerned him most was how these changes were directly linked to the changing nature of the urban land market and new networks of capital that realigned urban space and its economy beginning in the late nineteenth century.

In Mitra's writing new characters and sociopolitical networks represent the elemental changes that were under way within the urban economy. Mitra's Calcutta is a lifeless city dissected into plots in order to map out the vision of the urban developers. More importantly, *Agamikal* foregrounds that this new moral economy in the housing market was creating different networks of sociality.[17] Some of the central figures in his narrative therefore are land developers, engineers, accountants and technocrats who were redefining the city both materially, ecologically and ideologically:

> Daytime is measured by the tapes of the engineer, the road roller, the contractor's accounts, and songs of the coolies, yet under the omniscient moon, the mortified soul of the city wails.
>
> Perhaps progress and anguish are both real.
>
> The city is becoming independent. [And] someone is counting the loss of his fruit garden, someone's farmland – the tanks and ponds filled, palm, coconut and date trees felled – the city is progressing ... The big businesses look bloated. Two hundred new chimneys are reaching up to the sky to blacken it.[18]

The municipal response to this crisis remained focused on the missed or possible profits as a result of the changes in the urban land market. As we will see, new municipal regulations and a financial language emerged to tackle these "bloated businesses," either by creating exclusions within the housing market or through converting them into gainful investments. Swamp development and profit became central principles of urban planning, inaugurating an age of "residential capitalism."[19] By this time, Phelps was a well-known self-styled urbanist fighting for multiple causes, ranging from rationalizing footpaths and streamlining the existing parcel post system to remodeling the Town Hall of Calcutta. For Phelps and his supporters, mainly the British merchants of the city, the interwar period was a moment of "commercial boom" in the city.[20] Yet their grievance

[17] Recent studies on urban housing and neoliberalism have explored how moral economy of housing often creates new social networks; Caglar Keydar, "Globalization and Social Exclusion in Istanbul," *International Journal of Urban and Regional Research* 291 (March 2005): 124–34.

[18] "Diner alloy ingineerer gauj fite, roller, contractorer hishaab aar coolier gnaaiti, aar raater chaturdorshi chaander alloy tar pangu atmar ei kakuti. Hoyeto dui satya. Sahar shabalak hocche. Kaar gelo faler bagaan, kaar gelo fashaler kshet, gol patar gna uthlo – pukur dighi bharat holo, taal, narkel, khejur'er matha niulo – sahar egiye choleche ... saharer boro boro byaboshaiguli bujhi fnepe utheche. Dusho notun chimney uteche akasher mukhe kali makhate." Premendra Mitra, *Agamikal*, in *Premendramitra Rachanabali*, vol. 1 (Kolkata: Granthalay Pvt. Ltd., 1976), 225 (translation mine).

[19] Herman M. Schwartz and Leonard Seabrooke, "Varieties of Residential Capitalism in the International Political Economy: Old Welfare States and the New Politics of Housing," in *The Politics of Housing Booms and Busts*, ed. Leonard Seabrooke and Herman Schwartz (Palgrave: London, 2009), 1–27, 1.

[20] "Footpaths," *Amrita Bazar Patrika*, 14 November 1912, 10; "Parcel Posts: Conditions in Calcutta," *Amrita Bazar Patrika*, 31 January 1914, 7.

was that only the Marwaris partook in the profits, cheating the municipality and town committee out of their share of tax revenues.[21]

The soot-smeared chimneys that blackened the sky spoke to the industrial growth of Calcutta and its hinterlands. However, as Mitra's writing shows the developmental and commercial boom in the city signaled a moment of crisis and loss.[22] The jute industry, one of the largest in Calcutta, expanded from sixty-nine to eighty-five factories between 1914 and 1923, with a 36 percent increase in the workforce in Calcutta and its hinterlands.[23] While the 1920s marked the peak of prosperity for the jute industries of Calcutta, the city's working classes were suffering from problems of housing, health and chronic indebtedness.[24] Therefore, the developing city in Mitra's text was an anguished city, whose crisis was measured in the loss of gardens, ponds and farmlands and in the cries of the working class.

This urban crisis was connected with another figure, the land profiteer who enabled new circuits of money within the city, abetting the unbearable situation depicted above. The absence of solid land to expand the city became a culturalized phenomenon in the municipal and public debates. For instance, "Ditcher," a pseudonymic columnist of one of the most popular contemporary English dailies, *Amrita Bajar Patrika*, called the profiteers "house-bullies."[25] The "house-bullies," he asserted, were becoming a formidable presence in cities from Manchester to Calcutta. Moreover, according to Ditcher, they were aided in their speculative enterprise not by the absence of housing regulations but by the ignorance of the tenants about the existence of those laws. Drawing on his

[21] Rent Enquiry Committee, Municipal Miscellaneous Progs. §§34–45, April 1920, WBSA.

[22] From 1900, onward there was an increased growth of urban labor markets, especially in Calcutta and Bombay. Although the housing conditions were not ideal in the city, and laborers had to depend on agricultural income at home to sustain their families, nonetheless workers continued to come to the cities because of the loss of land, threat of famine, or whatever little chances of occupational mobility the urban labor market offered them. Tirthankar Roy, "Economic History and Modern India: Redefining the Link," *Journal of Economic Perspectives* 16, no. 3 (2002): 109–30.

[23] The percentage is computed using the data in Deshpande, *Report on an Enquiry into Conditions of Labour in Jute Mill Industry*, 6 and 28.

[24] Chakrabarty, *Rethinking Working-Class History*, 11.

[25] "Calcutta Rent Commission," *Amrita Bazar Patrika*, 11 December 1919, 6. There was a steady increase in joint stock holdings in land, estate and buildings in the Presidency towns from 1918 onward. See Subramanian, *Statistical Summary of the Social and Economic Trends in India (in the period 1918–1939)*, graph 25, 31. For a history of the new patterns of wealth accumulation and growth of private business in the Presidency towns in British India, see Bagchi, *Private Investment in India*.

experience with Manchester, he recommended to Phelps the immediate introduction of laws protecting tenants in order to check the rise of these "filibustering landlords," and thereby facilitate the improvement of the housing conditions in the city.[26] Yet, as we will see, the aggravated rental market was not simply a matter of regulating the activities of "filibustering landlords."

As elaborated in the previous chapter, new financial instruments related to taxation and management of urban land following the Crown takeover changed the economic geography of the city. By this time urban property had become a major source of income for those who could control its development. For the municipal authorities, rising property prices could become valuable sources of tax revenue. The government therefore sought its share of that income through increased taxation.[27] This in turn created the space for new modes of wealth accumulation and a new class of vernacular capitalists.[28] The expansive process of economic restructuring left an indelible mark on the urban geography of British India. One of the fallouts was the gradual process of financialization of the land market, which the war economy initially turned into a site of crisis followed by heightened regulation.

Within this colonial economic landscape, the flourishing jute mills and a robust growth in private investment in Calcutta significantly increased the presence of migrant workers and wealthy landlords alike. This also brought in substantial capital into the municipal coffers, which laid the financial groundwork for the establishment of the Calcutta Improvement Trust (CIT), almost twenty years after the Bombay Improvement Trust (BIT).[29] These colonial improvement trusts, intended as a solution to the

[26] "Calcutta Rent Commission," *Amrita Bazar Patrika*, 11 December 1919, 6.

[27] For an analysis of the parliamentary debates around India's budget, see Dacosta, *The Financial Situation in India, as Exposed in the Budget Statement for 1886–87*; for an in-depth analysis of the changes in the British Indian economy around the turn of the century, see Bagchi, *Colonialism and Indian Economy*; Birla, *Stages of Capital*; Dietmar Rothermund, *An Economic History of India: From Pre-Colonial Times to 1986* (London: CroomHelm, 1988).

[28] For a history of the emergence of vernacular capitalists and their modes of wealth accumulation, see Birla, *Stages of Capital*, and "Speculation Illicit and Complicit: Contract, Uncertainty, and Governmentality," *Comparative Studies in South Asia, Africa and the Middle East* 35, no. 3 (2015): 392–407; see also Hardgrove, *Community and Public Culture*.

[29] The bodies in charge of town planning through the nineteenth century, including Lottery Committees, Justices of Peace, City Magistrates and the Calcutta Municipal Corporation, were always hard-pressed for finances. It was only when the jute industry started booming that approval was given to levy duties on jute exported from Calcutta either

wide variety of problems facing industrialized cities, were derived from nineteenth-century town-planning experiments carried out in England and Scotland, whose specific form was determined very much by the prevailing political context and local elite groups.[30] In the case of both Bombay and later Calcutta, the idea was to have a separate agency undertake tasks of urban restructuring aided by "strong executive action unencumbered by accountability to representatives of local self-governing institutions."[31] What the CIT embodied was the centralization of particular legal instruments in a quasi-municipal body backed by the Bengal Legislative Council to acquire private property in the name of collective interest. In this manner, the CIT effectively fashioned itself as a land developer in the city from the moment of its founding in 1911.[32]

However, from the moment of founding, the CIT had to substantially distance itself from the BIT, its predecessor, established in December 1889, which had become an unpopular colonial institution. Founded initially as a response to the bubonic plague epidemic, the BIT was dominated primarily by colonial officials and Anglophone members of Bombay society.[33] In response to how the BIT dealt with the bubonic plague, the already disgruntled working-class population of Bombay

through land or sea. This export duty brought as much as 800,000 rupees annually around 1910s into the municipal coffers, which facilitated the establishment of the Calcutta Improvement Trust. Bompas, "The Work of the Calcutta Improvement Trust," 200.

30 For the case of Calcutta, see Datta, *Planning the City*. Perhaps no other improvement trust in India has received as much scholarly attention as the Bombay Improvement Trust. For scholarship on plague politics and the BIT, see Klein, "Urban Development and Death: Bombay City, 1870–1914"; Chandravarkar, *Imperial Power and Popular Politics*, 234–65; for a discussion of BIT as a failed example of urban planning, see Sandip Hazareesingh, "Colonial Modernism and the Flawed Paradigms of Urban Renewal: The Uneven Development of Bombay City 1900–1925," *Urban History* 28, no. 2 (2001): 235–55; Dossal, *Theatre of Conflict*; Kidambi, *Making of an Indian Metropolis*; and Frank Conlon, "Industrialization and the Housing Problem in Bombay, 1850–1940," in *Changing South Asia: Economy and Society*, ed. Kenneth Ballhatchet and David Taylor (London: Asia Research Service, 1984). The situation was similar in other colonial towns that were experimenting with improvement bodies; see for instance William Cunningham Bissell, "Conservation and the Colonial Past," in *Africa's Urban Past*, ed. David Anderson and R. Rathbone (Oxford: Oxford University Press, 2000), 240–61.

31 Kidambi, *Making of an Indian Metropolis*, 71–72.

32 P. J. Smith makes a somewhat similar argument about the improvement schemes of Edinburgh. See "Planning as Environmental Improvement: Slum Clearance in Victorian Edinburgh," in *The Rise of Modern Urban Planning*, ed. A. Suttcliffe (London: Mansell, 1980), 100–102.

33 Kidambi, "Nationalism and the City in Colonial India," 957.

mobilized themselves, threatening to spill over as riots in Calcutta.[34] By the end of the First World War the BIT occupied a tenuous position in the city, not just because the Bombay elite considered it completely inadequate to deal with the postwar housing situation, but also because the explosion of popular and working-class movements across Bombay demanded a reimagination of the role of urban development.[35]

Thus, it is no wonder that when in 1911 the Bengal Legislative Council passed the Calcutta Improvement Act, which established the CIT, the bill had a stormy passage in the Council. Most of the Indian councilors, drawing on the experience of Bombay, were wary of the drastic powers that the Trust might come to embody.[36] Yet the challenges that the new CIT faced, especially the housing shortage during and after the war, were very different from those of Bombay. While in Bombay large areas of government and municipal land were handed over to the trust with the anticipation that profit from improving the land, along with the annual municipal grant, would finance its improvement schemes, such bounty of land was nowhere to be found in Calcutta.[37] Swamp reclamation proved profitable for infrastructure, but that was not yet the case when undertaken for housing.

In this scenario, the wartime restrictions and demands had created a climate of profiteering in various essential goods like grains, cotton, jute and oil. Yet, given the finite nature of land in Calcutta, a different kind of governmental intervention into the land market was needed. This marks the peculiarity of land markets, for while a shortage in housing might be addressed through building on empty lands, something that was repeatedly proposed by both the British and the upper-class Bengali members of the municipality, the same does not always hold true for land.[38] Creating new land through law and draining was reaching its threshold in the twentieth century. Moreover, building expansion in a city like Calcutta during a time that Raj Bahadur Gupta called the period of "house famine" posed peculiar problems since the city was "shut up on one side

[34] Basu, "Strikes and Communal Riots in Calcutta," 954–61.

[35] By 1920 a new Development Directorate was set up in Bombay and the Trust was dissolved. See Kidambi, "Housing the Poor," 79.

[36] Bompass, "The Work of Calcutta Improvement Trust," 204.

[37] Bompass, "The Work of Calcutta Improvement Trust," 200; for a detailed account of BIT finances and land and housing schemes, see Kidambi, "Housing the Poor," 66–73.

[38] "Calcutta Housing: Government Criticism through Associated Press," *Times of India*, 18 June 1919, 7; "Calcutta News: City Development," *Times of India*, 4 June 1920, 18.

by the Hooghly [River] and the other by salt lakes."[39] The marshes of Calcutta emerged in the official discourses of the twentieth century by occupying a different register. They became limits to the development of property and profit. Overcoming this limit would require a different set of regulations and a collective amnesia that would enable a new visual literacy where marshes and water bodies would be regarded only as land-in-waiting. Moreover, to make matters worse, the cost of building and engineering material had also gone up during the war, which resulted in plummeting investment in construction as well as infrastructural works devoted to expanding transportation, tramline connections and bridges to the suburban parts of Calcutta.[40]

What role did the newly founded CIT play in abetting or aggravating the crisis situation? Regulation in the property market creates scarcity, rather than arising from it.[41] Instead of improving the conditions, the CIT further aggravated the situation, which directly contributed to creating a contracted housing market, on the one hand, and raising land prices, on the other.[42] The two schemes that impacted the land situation in Calcutta most were the "General Improvement Scheme" and the "Street Scheme." These were major public works involving significant land acquisition.[43] To enable these land acquisitions, the CIT established a special land acquisition collector (Valuer).[44] Unlike the situation in Bombay, the CIT had little access to the suburbs, so in 1913 it began the general improvement scheme by attempting to restructure the Burra Bazar, an area in central Calcutta, which had the highest concentration of masonry buildings and also happened to be an area where Marwari businesses were located.[45] While it was estimated that demolition of the unsanitary structures and rehousing would cost the CIT about 200,000 rupees,

[39] Raj Bahadur Gupta, *Labour and Housing in India* (Calcutta: Longman Green & Co., 1929), 159.

[40] Bompas, "The Work of Calcutta Improvement Trust," 209. E. P. Richards, the first town-planner of Calcutta, had proposed extensive tram lines for which the CIT began acquiring land from 1914. *Report on the Improvement and Town Planning of the City of Calcutta*, 105–20.

[41] Verdery makes a similar claim in her study of the transformation of the institution of property in post-socialist Romania. *Vanishing Hectare*, 16–17.

[42] In this its impact on the city was not very different from that of the BIT.

[43] Bompas, "The Work of the Calcutta Improvement Trust," 201.

[44] The board of trustees of the CIT was granted executive powers to acquire land through the 1894 Land Acquisition Act. *Calcutta Improvement Act, 1911*, Chap. IV, Sections 68–81, 346–60.

[45] As Bompas pointed out, this was an area covering almost ninety acres of land, with a population of 280 per acre. While the roads were better than the rest of Indian Calcutta,

C. H. Bompas, the first chairman of the CIT, pointed out that "the improvement effected was so great that good prices were obtained for the new building sites and instead of loss the re-sale of land resulted in a profit of rupees sixteen lakhs (Rs. 160,000)."[46] This was also a period when the CIT acquired a massive amount of land in this area. The income from land showed a steady increase, and after the war the CIT "obtained very high prices for the land which it sold and which it had acquired before the boom in all values began."[47] The profit was so great that, instead of the annual estimated return of 100,000 rupees, the Trust made 1,200,000 rupees.[48] By 1922, apart from benefiting from the lucrative land market (partially of its own making), the CIT had displaced more than 50,000 people. At this point the CIT also had in its hand land estimated to be worth £1,100,000 in Calcutta.[49] It is no surprise then that some of the nationalist members of the municipal commission sarcastically designated it as the "body which is mainly responsible for dehousing of the people [of Calcutta]."[50] While the rent debates never mention CIT's role in this light, the work of the CIT radically changed the relation between land price and the rental market in the city. Then what were the members of Rent Commission saying?

From the perspective of the municipal officials, traders, administrators and town planners, the congestion of Calcutta was a problem that they could bring under control only through effective road and transportation planning and rationalization of the land market. According to them a proper and affordable communication system would enable the working-class population to move out of the city and thereby clear the city of slums. Accelerated public works, displacement, and coercive processes of land acquisition were among the many causes for land and rent prices skyrocketing, which permanently restructured the urban geography of the city.[51] Yet both official and public opinion held that something else was

amounting to 16 percent of the total area, given that it was a business area and congested he proposed his improvement scheme on the Marwari-owned Burra Bazar area.

46 Bompas, "The Work of the Calcutta Improvement Trust," 206.

47 Bompas, "The Work of the Calcutta Improvement Trust," 209.

48 Bompas, "The Work of the Calcutta Improvement Trust," 210.

49 Bompas, "The Work of the Calcutta Improvement Trust," 212.

50 "An Extract from an Abstract of the Proceedings of a Meeting of the Bengal Legislative Council Held on Tuesday, April 13, 1920," 34. IOR/L/PJ/6/763.

51 Kidambi shows how, instead of alleviating the housing shortage, the BIT's work increased homelessness, creating a shortage in housing and a spike in the rental market in Bombay, which ended in the 1919 riots against the extension of the war restrictions. Kidambi, *Making of an Indian Metropolis.*

steering the movement of prices within the land and housing markets. Thus, on 3 September 1919, Phelps proposed a motion that required the municipal commissioners to turn their attention to the housing market as tenants were being "squeezed" out of their means and being forced to pay extortionist house rents. This, according to him, was because of the new phenomenon of hoarding land without investing in building activity. "Squeezed," as we shall see in this chapter, emerged as concept bandied about in municipal debates to express the workings of Calcutta's housing market. Such "squeezing" was a clear sign of a contracted housing market. Phelps and the British merchants demanded a particular relation between municipal governance and the housing market, a relation where the commissioner, learning the laws of this market, would finally regulate its vicissitudes. Following the motion, the Rent Commission was established and presided over by Phelps.

The official debates on how to curb housing speculation during the 1920s can be read as an early bureaucratic exercise in rationalizing the urban land market and making it efficient by regulating the ways in which land and housing accrued value economically. It also demonstrated how landlords got typecast culturally. Studying the exclusions that were enframed in the early twentieth-century housing market reveals what was unique to this moment.[52] Phelps and his team of British merchants were not aghast at speculation as a corrupting element that had entered the domain of the housing market and political life in the city. They were much more concerned with how to profit from it through market-based speculation by curbing what they understood as pathological profiteering.

A NUMERICAL CONTROVERSY?

The Calcutta Municipal Corporation established the 1919 Rent Commission with "the sole purpose of investigating the sudden abnormal rise in property rent."[53] The commission was confounded by a "curious tendency," as they noted: "The desire to invest in landed property has also an

[52] Enframing, according to Timothy Mitchell, is about the creation of a set of techniques, arrangements, practices of hierarchies and classificatory regimes to know something. In the case of a market, he argues that enframing must also include a study of the "exclusions on which market transactions depend." Timonthy Mitchell, "The Properties of Market," in *Do Economists Make Markets? On the Performativity of Economics*, ed. Donald Mackenzie, Fabian Muniesa and Lucia Siu (Princeton, NJ: Princeton University Press, 2007), 245.

[53] Rent Enquiry Committee, Municipal Miscellaneous Prog. April 1920, §1, WBSA.

accumulative effect; the contagion spreads, and instead of the high prices being limited to property, which is capable of increasing very greatly in value in the near future, speculative prices are being paid for *any property* coming on the market irrespective of its possibilities."[54] Rangoon's Land Ceiling Act enacted in 1918 became the template for understanding urban property issues in Calcutta.[55] According to the Rent Commission, "The papers which [they] ha[d] obtained from Rangoon disclose the interesting fact that a rise very similar to the one which we are dealing with manifested itself in that city from about May 1919."[56] While Calcutta had its fair share of working-class mobilization since the 1890s around issues of plague regulations, wages and holidays, it was the riots in the working-class neighborhoods in Bombay in 1918, protesting war restrictions, that provided those sitting in Calcutta with a cautionary note about what the mismanagement of the property regime by the colonial government could foment.[57]

Phelps was concerned that, since the outbreak First World War, house rents had doubled in a mere five years. Some members of the Legislative Council tried to point out that perhaps the reasons for rent escalation were related to the increased costs of building materials, which had become expensive during that period.[58] Phelps, however, was not convinced by that argument. Instead, he raised doubts that a rise in the cost of building materials could be so significant that it would justify the "doubling of rent of houses which were erected before the war disturbed current prices."[59]

Phelps's doubts were further solidified by the deposition of one of his staunchest Indian supporters, the Maharaja of Burdwan, who had a considerable number of rental units in Calcutta. He concurred with Phelps that the housing market was inflated.[60] He pointed out that, as a

[54] Rent Enquiry Committee, Municipal Miscellaneous Prog. April 1920, §4, WBSA.

[55] Rent Enquiry Committee, Municipal Miscellaneous Prog. April 1920, §§2–3, WBSA.

[56] Rent Enquiry Committee, Municipal Miscellaneous Prog. April 1920, §2, WBSA.

[57] In a speech by the Honorable Surendra Nath Banerjea introducing the bill to amend the Calcutta Municipal Act, 1925, he talks about working-class riots. The riot discussed in this speech refers to the 1919 protest held in Bombay in response to the colonial state's decision to extend wartime restrictions on civil liberties. See Basu, "Strikes and Communal Riots in Calcutta"; and Kidambi, "Nationalism and the City," 960.

[58] Bompas, "The Work of the Calcutta Improvement Trust," 212.

[59] Rent Enquiry Committee, Municipal Miscellaneous Prog. April 1920, §7, WBSA.

[60] The Maharaja of Burdwan was among those English-educated Indian publicists and part of the landed gentry who were very active within the municipal and civic affairs of the city who often adapted ideas from British liberal vocabulary, especially notions of

Bengali landlord operating in Calcutta, he was aware of the "rapacity of certain landlords" who were new in the market. According to him, there was a general rise of rents in units under the control of these new landlords, especially in the Burra Bazar area, as well as a rise in sale prices of houses across Calcutta. Of course, neither Phelps nor the Maharaja brought up the "rapacity" of the CIT's activities in Burra Bazar before and after the war. The Maharaja of Burdwan claimed that a property that was priced at 380,000 rupees in 1918 changed hands two years later in 1920 for 2,200,000 rupees. He went on to add that, if three-bedroom rentals primarily for middle-class and English residents in Lower Circular Road, or Elliot Road and Park Street, witnessed a doubling of rental prices, then one-bedroom working-class housing in areas like Zakaria Street or Boloram De Street were not spared either. On the one hand, he admitted that some of the Indian landlords might regard governmental intervention as interference in the private rights of all landlords in Calcutta. Yet, on the other hand, he admitted that in a case like the one in front of the municipal council "where profiteering becomes a general and flagrant abuse, ... state control in [the] economic sphere [could] be most easily justified."[61]

Arguments like this firmly established a causal relation between the housing crisis and the entry of new upcountry and Marwari landlords in Calcutta's land market.[62] The 1919 Rent Commission report argued that World War I slowed down building activity in Calcutta, even while its population increased considerably. At the same time, they pointed out that, as a result of the relocation of major European businesses to Calcutta, the North Indian commercial classes began augmenting their accumulation of wealth quite significantly. The period of armistice was thus

improvement, public good and progress. These ideas often became very much entrenched as a particular kind of Indian civic culture. See for instance Anderson, "Public Nuisance and Private Purpose." Similar arguments have been made for other cities too. For Bombay, see Kidambi, *Making of an Indian Metropolis*; Hazareesingh, *The Colonial City and the Challenge of Modernity*; see also Haynes's study of Surat, *Rhetoric and Ritual in Colonial India*.

61 "Minutes of Meeting at Calcutta Town Hall, March 15, 1920," 1–6, IOR/L/PJ/6/763.

62 Rent Enquiry Committee, Municipal Miscellaneous Prog. April 1920, §§4–6, WBSA; "Land Values and House Rents: Advantages of a Leasehold," *Amrita Bajar Patrika*, 2 October 1919; "Land Values and House Rents: Some Restrictions," *Amrita Bajar Patrika*, 19 September 1919.

marked by a "permanent lack of equilibrium in the demand and supply of the housing market in Calcutta,"[63] just as in the grain, piece-goods and commodities markets.[64] This was caused by the fact that landlords were hoarding land instead of selling land and investing the profit into new channels that would aid the market. As we will see later, one commendable investment was the project of draining the marshes of eastern Calcutta to develop the land for housing. In short, a good landlord also had to be a good land developer.

Increased commercial activity had a perceptible effect on the housing and the rental market in the city. Though many causes were cited in the municipal and town hall debates, it might be worthwhile to analyze the manner of bureaucratic reasoning through which speculation was placed squarely on wealth accumulation among the Marwaris.[65] Phelps's report detailed that, from 1909–10 to 1913–14, the Calcutta Corporation sanctioned 1,702 masonry buildings. This was slightly higher than the number sanctioned during the war years, 1914–19, when the number was 1,628. It became clear that urban property speculation was an "epidemic" that was spreading through the city because of the "unreasonableness developed by a few landlords."[66] The Rent Commission argued that the rise in rent was so steady that it warranted "the theory of a five-percent annual increase," which generally signified a city growing in wealth and health. This report, however, did not explain why a 5 percent steady rise in rent signaled a city growing in "wealth and in health."[67] Moreover, this image of a healthy and wealthy city runs counter to Phelps's central claim of the squeezed tenants comprising both European officials and the Indian working classes caused by the unhealthy market practices of the extortionist landlords.

Interestingly, Phelps substantiated his "theory of a 5-percent annual increase" in housing rent not through numbers drawn from the housing situation of Calcutta, but rather by furnishing statistical documentation on the rise in cost of living in various Presidency towns in India.[68]

63 "Report of the Calcutta Housing and Communications Committee 1923," 1, WBSA.
64 Bagchi, "Indian Economy and Society during World War One."
65 One of the curious explanations among them was the sudden emergence of Japanese landlords in a Bhawanipore area, which was hardly entertained. Municipal Miscellaneous Prog. April 1920, §§4–6, WBSA.
66 Rent Enquiry Committee, Municipal Miscellaneous Prog. April 1920, §6, WBSA.
67 Rent Enquiry Committee, Municipal Miscellaneous Prog. April 1920, §3, WBSA.
68 For an account of the performance of statistics in representing a problem of governance and policy-making, see Emmanuel Didier, "Do Statistics 'Perform' the Economy?," in *Do*

The little documentary evidence we have on land prices in Calcutta from that period show that the price of suburban lands outside Calcutta, primarily the wetlands, was rather high compared with that of central Calcutta. Mr. C. E. Payne, the then deputy chairman of the Calcutta Corporation, pointed out that it was impossible to buy suburban land in Calcutta, unless it was an undrained swamp costing almost £4,500 per acre as compared with £200 to £300 per acre for similar undeveloped land in England.[69] More than any other factor, the price of land made expansion a costly affair.[70] According to the Rent Commission, the rising price of essential goods depreciated people's purchasing power and contributed directly to the rise in land value. Citing the index numbers of retail prices of food and essential goods from the Department of Statistics, the Commission argued that

> it would be easy to advance theories of doubtful soundness in regard to this point [of the rising cost of living], but when the prices of the necessaries of life are found to have increased 54 per cent in the last five years, and it is seen that people generally have to pay more for everything than formerly, the temptation to the land-owner to adjust his income to his increased cost of living is no doubt a factor in his mind when fixing rents.[71]

While numerical data on the rising cost of food, coal, cotton and essential goods can perhaps provide evidence for the rising cost of living (see Figure 5.1), it can hardly be the basis for forging the "fact" of land-profiteering by unscrupulous (Marwari) landowners in Calcutta. Yet that was exactly what the municipal discussion brought forth. Indeed, there was an enormous amount of statistical data on price movements from the various corners of the colony for the past half-century, which Phelps mined and marshaled for his purposes to rationalize the housing market in Calcutta.[72] Yet, if we turn to the graphs of rising prices shown in

Economists Make Markets? On the Performativity of Economics, ed. Donald Mackenzie, Fabian Muniesa and Lucia Siu (Princeton, NJ: Princeton University Press, 2007), 276–310, 277–79.

69 Bompas, "The Work of the Calcutta Improvement Trust," 214.

70 It is not surprising perhaps, that land in this very same area (Salt Lake in contemporary Kolkata), about which Payne and Bompas expressed puzzlement, fetches an exorbitant price.

71 Rent Enquiry Committee, Municipal Miscellaneous Prog. April 1920, §10, WBSA (emphasis mine). The report tabulates the rising cost of essential goods, which I have not reproduced here. Housing is not part of the tabulation.

72 In 1862, the British Raj set up a Statistical Committee for administrative purposes and by 1869 the Director General was producing comprehensive statistical surveys of the country. For a very brief survey, see S. Subramanian, "A Brief History of the Organisation of

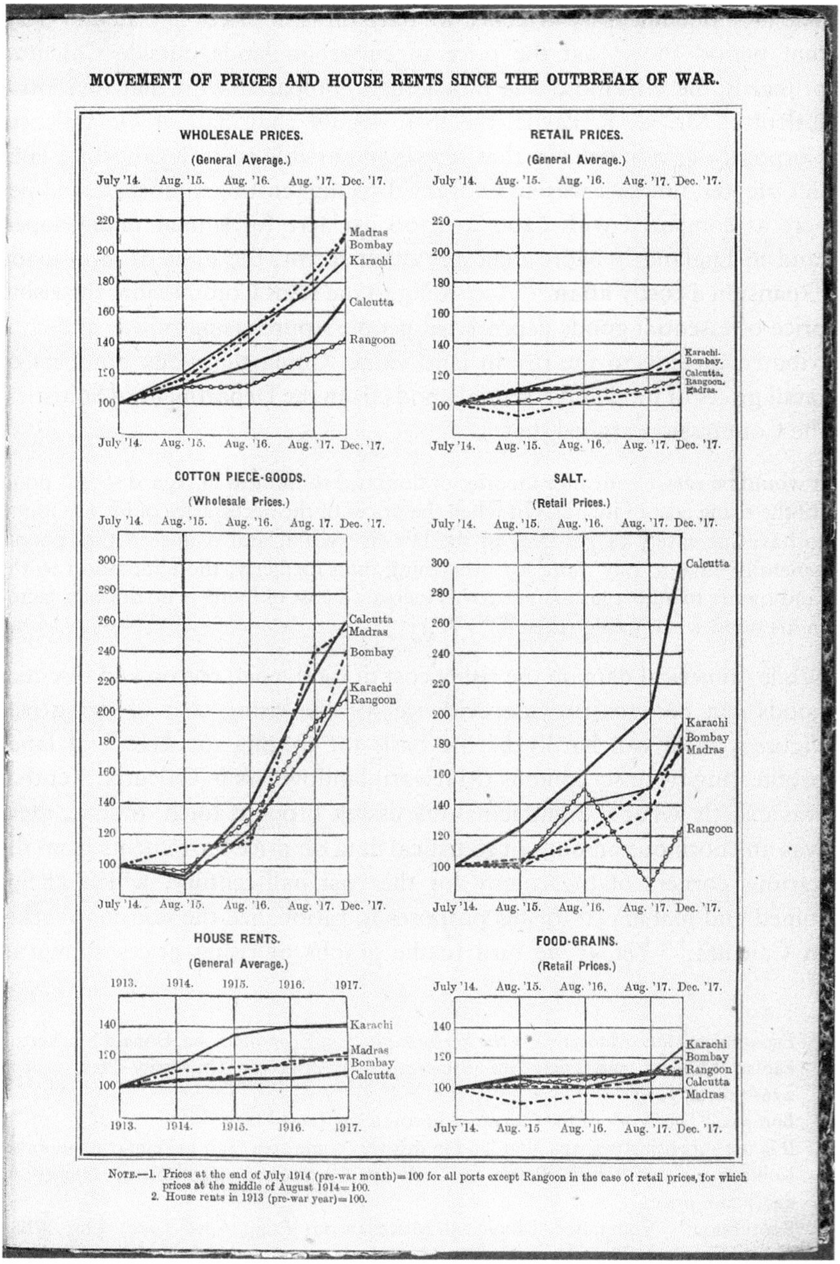

FIGURE 5.1 War prices and house rents. Statistics showing wholesale and retail prices and house rents at the end of December 1917, as compared with the prewar level (July 1914), August 1915, 1916, and 1917 (Department of Statistics: Calcutta, 1918).
With permission from Harvard University Library.

Figure 5.1 published just two years earlier (and that surprisingly did not become part of the report), they tell a rather different story of property prices in Calcutta than what Phelps's report, the Rent Commission debates or even the public outcry illuminate.

Surprisingly, Figure 5.1 shows that the cost of cotton, food grains and salt increased greatly in Calcutta in comparison to other cities like Karachi or Bombay. However, what is important is that the figure also highlights that Calcutta experienced a much lower rise in rent compared with the other cities. If these graphs reveal anything, it seems that there is no direct correlation between the rising cost of necessaries for living and the landlords' predilection to fix a higher rent to supplement their income. The Department of Statistics claimed that the tables were not without indeterminacy and must be read as a "finger post of the fluctuation of price levels ... These index numbers do not pretend to show the relative prices in different ports or provinces since the articles are, for example in the case of Calcutta, not entirely the same as elsewhere."[73]

Relative or not, no one can deny that the cost of living was rising in Calcutta. The nationalist movement was gathering momentum in the city and the situation was not quite under the control of the colonial officials. Heightened working-class mobilization had forced the city commissioners and factory owners since the turn of the century to debate who was responsible for providing the factory workers with housing.[74] A mass gathering was held in College Square hours before the town hall meeting of 3 March 1920 where Phelps proposed a solution to the housing crisis in the form of a Calcutta Rent Bill. Nationalist leaders, labor union leaders and mill, dock and factory workers attended the mass meeting. J. N. Roy, the barrister and president of the Central Labour Federation, who had socialist sympathies, gave a fiery speech at the meeting denouncing the "Marwari speculators and up-country profiteers" as responsible for the increase in housing prices.[75] There is no evidence to account for how Roy came to this conclusion. However, such ethnicization of the housing market by the Bengali population of the city is not surprising given the

Official Statistics in India during the British Period," *Sankhya: The Indian Journal of Statistics* 22, nos. 1–2 (1960).

73 "War Prices and House Rents," 1.

74 Chakrabarty, *Rethinking Working Class History*.

75 "Calcutta Rent Problem," *Amrita Bazar Patrika*, 12 March 1920, 8.

heightened colonial regulation of vernacular economic practices from the turn of the twentieth century.[76]

The commissioners sitting in Calcutta, including Phelps and his team, garnered statistical data accumulated from various sources to produce the *fact* of "profiteering" and "crisis." During the interwar period, in many parts of the world statistical bureaus were being called on by the governing officials to apply their numerical knowledge to combat what was known as the "ravages of speculation," since statistical objectivity would prevent distortions in the markets.[77] The 1919 report displayed a similar symptom of viewing numerical representations as incontrovertible facts that would manage this volatile housing market. The 1919 report adapted the earlier epidemiological tropes of urban governance as crisis management to the new fascination with numerical objectivity. Yet, as we will later see, in spite of their good intentions, many of the nationalist members of the team were upset with Phelps, since this objectivity failed to distinguish the Bengali, European, Anglo-Indian or the Armenian landlords from the Marwaris.[78]

After citing the malarial hinterlands and the absence of arterial roads connecting the city to the cheaper lands of the suburbs as reasons for the rising rent, the 1919 report stated the following:

> The innate unwillingness of certain classes of the population to leave the area, in which their ancestors have lived for generations, and the immobility of certain classes of the population, while perhaps not major factors affecting the whole city, have undoubtedly contributed to high prices for landed property and high rents in the northern sections.[79]

The report employed arguments drawn both from the field of public health and from assumptions about cultural difference typical of colonial government's representational language and adapted these arguments to the statistical data: numbers in the service of colonial racism. Perhaps we need to look carefully at how the language of economic difference constructed one particular community and its landholding as a site of governmental intervention.

[76] Birla, *Stages of Capital.* [77] Didier, "Do Statistics 'Perform'?," 278

[78] "An Extract from an Abstract of the Proceedings of a Meeting of the Bengal Legislative Council Held on Tuesday, April 13, 1920," 64. IOR/L/PJ/6/763.

[79] Rent Enquiry Committee, Municipal Miscellaneous Prog. April 1920, §3, WBSA.

Statistical indices were producing knowledge about costs of the standard of living.[80] In Phelps's opinion, it was the duty of the administrators to legislate according to those statistical facts about living costs and the market supply. Although rent was under the municipal domain, Phelps wanted more intricate economic-legal measures in place to control rent and protect tenants.[81] He noted with dismay:

> The most astonishing objection was, in my opinion, that the matter of rents was entirely and very properly subject to the laws of supply and demand. The commissioners professed their inability to see the difference between the purchase of a pair of boots and the renting of a house.[82]

The commissioners failed to understand that, although the housing market was operating through its own set of laws, it could nonetheless be subjugated to the domain of judicial law. This failure prevented Phelps from carrying out his wishes to reorder the colonial housing landscape through law. According to Phelps, the market and its laws could and must be controlled by the judiciary:

> The law of supply and demand became a fetish, a law which might sometimes act cruelly but must nevertheless be grovelled to with hopeless adoration. But the same people [municipal commissioners] voted for the control of the "'common necessities of life'" such as coal, oil, *ghee* [clarified butter], rice and cotton-piece goods ... I ask ... is not a house, a flat, a dwelling place of some sort a common necessity of life?[83]

His perspicacious contention was that, unlike boots as an item of luxury, housing and urban property had always been common necessities and were now governed by market logics of supply and demand, which currently were in a state of disequilibrium. Phelps demanded a refined understanding of the market laws in the service of urban governance. For the rest of the members consideration of market behavior was still outside the domain of public administration. Phelps was enterprising enough to understand that the market order had to be folded into the legal order, if the colonial coffers were to remain full and funded.

Phelps was not the lone voice. Other voices of support, albeit with a different sense of purpose, came from the British and European merchants

80 Didier, "Do Statistic 'Perform'?," 213.

81 Although he was not successful initially, the Court of Appeals to adjudicate housing issues was again brought to the table of the Legislative Council and Town Hall meetings. Report of the Calcutta Housing and Communications Committee 1923, §16, WBSA.

82 Rent Enquiry Committee, Municipal Miscellaneous Prog. §10, April 1920, WBSA.

83 Rent Enquiry Committee, Municipal Miscellaneous Prog. §10, April 1920, WBSA.

and tradesmen in the city. Although for a long time the nonofficial British merchants and traders in this Eastern colony did not overtly interfere in administrative matters, by the early twentieth century the nonofficial British community in Calcutta numbered almost 23,000 and wielded significant influence in the political debates.[84] Moreover, with the flourishing of the jute and tea industries in the early twentieth century, these two commercial organizations became important intermediaries with which the British governing bodies had to contend.[85]

Housing was becoming increasingly precious to the British with every passing day. What made matters grave was that new wealth accumulation among the Marwaris created new networks of power and a new moral economy in the housing market. Older forms of loyalties and negotiations failed to work in this new setting. In this situation British officials found it hard to rent back their earlier properties when they returned to Calcutta from their summer capital in the Himalayan town of Shimla every October. Either the property was already rented to someone else or it was available at a newly negotiated rent.[86] Although rental units were in higher demand than the available supply, the emergence of a new class of wealthy upcountry merchants meant that the British were seeing a withering away of their powers as both landlords and tenants within Calcutta's rental market.

By the 1920s the British merchant community decided to set up a housing trust, which would act as a recommending agency to the government on the housing crisis in Calcutta. A. D. Pickford, a tea merchant and member of the Indian Legislative Council, served as the secretary of the Calcutta Domiciled Community Enquiry Committee and wrote to the secretary of the Government of Bengal requesting land to set up a housing trust, following up with "a strongly worded report on the conditions of housing in Calcutta so far as this particular community [nonofficial British community] [wa]s concerned."[87] He stated that he had to

[84] Broomfield, *Elite Conflict in Plural Society*, 43. Rudyard Kipling pointed out how the British population of India around the turn of the century was not a homogenous group but a rather stratified collection of people divided according to trade and affiliation to the Raj. "Bengal Civilians," "Government of India Men," "Men of the Firms" and the "Tradesmen," all in *The Phantom Rickshaw, City of Dreadful Night and Other Tales* (New York, NY: Lovell Company, 1890), 13.

[85] On the importance of jute, tea and the paper industry and the emergence of private investment around this time, see Bagchi, *Private Investment in British India*.

[86] Papers Relating to the Calcutta Rent Bill, 64–65, IOR/L/PJ/6/763.

[87] Proposal for the Acquisition of Land under the Land Acquisition Act on Behalf of a Building Trust for Housing the Anglo Indian Community in Calcutta, Government of

approach the government to acquire lands for the purposes of the Trust building since the "wild speculation in land" made it difficult for them to acquire land at a reasonable price. Speculation, coupled with the manner in which land was owned in small parcels in Calcutta, necessitated government intervention. In their logic, controlling speculation should be materialized (literally) on the ground by rationalizing landholding. According to the Trust, too much land was held *extra commercium* by unscrupulous North Indian landlords. Those lands and the profits from the burgeoning rental market had to be brought into circulation in order to stabilize the market.[88] Those profits could in theory be channeled to drain the marshes surrounding Calcutta. Their request for subsidized land to equalize the land market went unheeded, since there was no provision in the Land Acquisition Act of 1911 to acquire lands for private purposes or for the private interests of any merchants.[89]

Ecological risk management had been supplanted by financial risk governance in the land market. The domiciled European and Anglo-Indian community, as well as the established Bengali landed gentry, were considered "the investors and developers of land."[90] Following the commercial boom in the city between 1917 and 1919, when the market price of land went up, the Anglo-Indian and Bengali landowners too indulged in both speculative buying and the selling of urban land when the price was conducive. However, according to popular discourse, unlike the Marwari landlords, they did not hold onto land, against supposedly sound market principles.

It turns out, then, that land-profiteering was not new. The problem with twentieth-century profiteering lay elsewhere: in the art of speculating.[91] The municipal officials, following the symptoms of the market in urban land, pointed out that the lessons learned through calculated speculation could dovetail into market rationalization, and the so-called speculative buyer could enter the market as a seller:

Past history has proved that prices paid for land in Calcutta, which at the time were considered to be of a speculative character, have soon been overtaken by the

Bengal Proceedings, Municipal [Miscellaneous] Department Prog. 13–15, March 1920, WBSA.

88 For a critical reading of the myth that untitled land is *extra commericium* or "dead capital," see Mitchell, "The Properties of Market," 248–49.

89 Reply by Mr. O'Malley, Secretary to the Government of Bengal to A. D. Pickford, Municipal Miscellaneous Prog. March 1920, WBSA.

90 Report of the Calcutta Housing and Communications Committee 1923, §11, WBSA.

91 Birla, "Speculation Illicit and Complicit," 392–407.

market, and the so-called speculative buyer within a comparatively short period of years has been able to sell at a profit. This historical fact is well known to the persons who are now paying what appear to be high speculative prices; and this, combined with the large amount of free money and the indications of a commercial boom in Calcutta, has no doubt led to the payment of very high prices.[92]

In their logic, speculative buying when followed through with a well-timed sale of lands would allow the market to regulate housing cost. However, they could not decipher what financial principles made the Marwaris hold onto land and not sell it even when profits might be made. Selling at a profit would improve the supply of housing in the market, as well as the general market conditions. Therefore, they posited, the Marwaris fixed rental prices on a whim, instead of letting rents be regulated and determined by a healthy demand and supply within the housing market.

Rent, according to the findings of the report, must be determined scientifically through a set of calculations based on principles of exchange in order to protect the interests of both the landlord and the tenant. This would be called "economic rent," as opposed to the current "extortionist rents" that were being charged in many parts of Calcutta.[93] The rent debates proposed three kinds of rents. Economic rents were generated automatically by the market. Standard rents were artificially generated through a set of restrictions and controls and were based on calculating what the tenants can pay. Standard rents would help equalize the market, which would then be able to generate economic rents. Extortionist rents were what the Calcutta landlords were charging their tenants.

The principle of rent determination was not supposed to be contingent on either lengthy negotiations or the whims of the landlord. Rent could become a reliable piece of economic information only when it was produced through healthy market competition. Within this scenario, speculative buying and selling were permissible when it made the vicissitudes of the housing and rental market a knowable entity in order to generate the so-called economic rent.[94] So a rational speculator was supposed to build on prior lessons to enter the market with reliable economic information, and not flow with the vagaries of chance, which was demarcated as the domain of the native profiteer.

[92] Rent Enquiry Committee, Municipal Miscellaneous Prog. April 1920, §3, WBSA.
[93] Papers Relating to the Calcutta Rent Bill, 50–2, and passim, IOR/L/PJ/6/763.
[94] Rent Enquiry Committee, Municipal Miscellaneous Prog. April 1920, §§17–20, WBSA; Report of the Calcutta Housing and Communications Committee 1923, §20, WBSA.

According to the report of 1919, this "particular lesson of the past history" unfortunately proved disastrous for the urban market, as "certain commercial classes" were now holding onto the land bought at higher prices and refusing to sell it even when profit would accrue.[95] The colonial officials failed to understand the indigenous commercial classes as future speculators turned market agents; they were instead regarded as merely hoarding land and contracting the market. Thus, for the British officials, capriciousness, unlike rational exchange, played a dominant role in commercial transactions undertaken by the indigenous population. This sort of behavior could only be read as a symptom of a larger incapability on the part of the native inhabitants to see the future in terms of economic development.

As the later Land Development Report from 1923 accused, landowners often refused to sell land not because they could not get higher prices, but because of their purported inability to make sound decisions, which stemmed from "absence of sound ideas of land finance."[96] The Land Development Report of 1923 also pointed out that the main problem was the disproportionately exorbitant cost of land versus building material in Calcutta compared with London or other cities of England.[97] The cost of land in housing schemes in England between the years 1919 and 1922 averaged out to 50 rupees per kottha (approximately 720 square feet), which is "less than a tenth of Calcutta land values in the outer residential areas."[98] Land had therefore emerged as a highly sought-after urban commodity. The report translated this numerical deduction into a racial pathology for holding onto land. This pathology was presented through the following analysis offered by the report:

The Committee have examined the reasons for the high cost of land and analyse them as follows:

a) The great demand for land ownership.
b) Unwillingness to sell at less than a rate in the owner's mind. He prefers to wait until he "gets his price."
c) Complete absence of any conception of loss of economic interest on capital invested in land.
d) Satisfaction for the great body of landholders with a low rate of interest (on their idea of capital value) in the form of rent.
e) The attractiveness of landholding as a form of investment for safety.

95 Rent Enquiry Committee, Municipal Miscellaneous Prog. April 1920, §3, WBSA.
96 Report of the Calcutta Housing and Communications Committee 1923, §15, WBSA.
97 This is also corroborated in Bompas's account.
98 Report of the Calcutta Housing and Communications Committee 1923, §6, WBSA.

f) The opportunity that landholding gives for gradual investment by way of building development. *Building is almost a national recreation in Bengal.*
g) The desire to provide housing for descendants.
h) The tendency of the Indian family system to keep relatives in one locality and so produce inter-acting pressure.[99]

Economic maneuvering conducted in a rational manner produced "economic rents," but the Housing Committee located native practices of investing in land within a cultural space of ignorance.[100] The 1923 report suggested a remedy, and that was "to educate the Calcutta [indigenous] public in the elements of land finance."[101]

Speculation was a professional activity, while profiteering was portrayed as illogical mimicry without professional knowledge or understanding of exchange. Such hoarding of land produced a contracted rental market, which merely facilitated one form of wealth accumulation, rather than circulation and investment in future land development projects. As far as the colonial officials were concerned, what the Marwari landlords were doing was not efficient maneuvering of the market based on deliberate calculation. Therefore, it was the imperative of the municipality to introduce regulation in the form of "standard rents," which would be put in place for three years to bring equilibrium to the housing and rental market that could then automatically generate "economic rents."[102]

In a series of editorials written for the *Amrita Bajar Patrika*, Phelps gave free rein to his ideas of restructuring the municipal administration of urban land. His prognosis was that "what we are suffering from in Calcutta is a shortage of houses, not-withstanding that there is plenty of land on which to build, but that the unimaginative people who have money are actuated rather by a desire to increase their wealth than to use it in a *reasonable* and *wholesome* direction to improve the city and to provide accommodation for the working classes."[103] His answer to this problem was to propose a new Calcutta Rent Bill.

Apart from a call to reassess property values annually instead of every five years, Phelps wanted to introduce a three-tier system of regulation to curb the rampant housing and urban property speculation in early

99 Report of the Calcutta Housing and Communications Committee 1923, §6, WBSA.
100 Papers Relating to the Calcutta Rent Bill, 50, IOR/L/PJ/6/763.
101 Report of the Calcutta Housing and Communications Committee 1923, §24, WBSA.
102 Papers Relating to the Calcutta Rent Bill, 68, IOR/L/PJ/6/763.
103 "Land Values and House Rents: Suggested Restrictions," *Amrita Bajar Patrika*, 19 September 1919 (emphasis mine).

twentieth-century Calcutta. First, he wanted to extend the wartime regulation to the housing market. According to Phelps, such a regulation had to be modeled on a similar measure introduced in England. The English Rent Act of 1915, "a purely war measure passed for the duration of the war and six months after," was extended to the year 1921, even though it was supposed to be implemented only during the war years. Moreover, if the English municipal commissioners considered any property to be of very high rental value, they could pass an order bringing those houses under the legislative control of the Rent Act of 1915. Such expansive legislative powers were not available at the municipal level in British India. Although the 1919 Rent Act of Bombay charted a similar course to the British act and was extended for another two years, the same did not immediately happen for Calcutta, in spite of Phelps's efforts.[104] A considerable portion of the Bengali community, who were rate-paying members of the municipal commission and landowners, vehemently opposed the proposal for a rent bill, finding it not only an unjust price they had to pay for the "rapacity" of the North Indian commercial classes, but also a measure that would discourage investment in building new rental units.[105]

Second, Phelps demanded secure laws to protect tenants from unscrupulous landlords in the form of a court of appeals. In order to ensure the implementation of judicial review of tenant–landlord disputes over rents, he proposed his new Calcutta Rent Bill.[106] Finally, he demanded the need for a financial climate that would encourage private investment in developing land without state intervention. His vision of private investment was very distinct. He upheld the example of the work done in the late nineteenth century by the real estate manager and Armenian philanthropist Mr. J. C. Galstaun in developing the working-class neighborhoods in the prime area of Camac Street, a predominantly British part of colonial Calcutta.[107]

[104] Rent Enquiry Committee, Municipal Miscellaneous Prog. April 1920, §1, WBSA.

[105] "An Extract from an Abstract of the Proceedings of a Meeting of the Bengal Legislative Council Held on Tuesday, April 13, 1920," 42–67. IOR/L/PJ/6/763.

[106] In an editorial published in October 1919, Phelps explicitly mentions that Marwari, along with some Armenian and Jewish landlords in Calcutta, made lives less secure for the tenants for both the working classes and the British community. "Land Values and House Rents: Advantages of a Leasehold," *Amrita Bajar Patrika*, 2 October 1919.

[107] "Land Values and House Rents: Some Restrictions," *Amrita Bajar Patrika*, 19 September 1919. Mr. J. C. Galstaun built many mansions and endowed many parks in Calcutta, devoted as he was to projects of beautification. He worked extremely closely with the British officials in Calcutta. For a brief biography of Mr. Galstaun, see Mesrovb Jacob,

In short, Phelps's recommendations envisioned a restructuring of the relation between the municipal governance, the land market and private capital. If more powers were granted to "capitalists," it would result in more investment in draining the suburban swamps and building activity in the city.[108] Such investments in land and building would be the financial instrument that would increase the quantity of housing to meet the heightened demand. Draining the swamps would create more land, thereby bringing market equilibrium. This would also curb the excessive power that the hoarding-prone landowners now wielded in Calcutta's rental market.[109] Controlling ecological variability through draining would serve two purposes. First, it would create a profitable market, and second, it would impede the boundless wealth accumulation of the Marwari landlords.

Such undifferentiated characterization of the landowners did not go down well in the Rent Bill meetings that preoccupied the municipal debates through most of the 1920s. The bill was hotly debated on the municipal and town hall floors for a year and a half before being passed on a much-compromised scale. While every member, British, European, Anglo-Indian and Bengali, agreed on the need to curb Marwari speculation, they argued over the details of the bill and how best to protect their own community's interest. Bengali members like P. C. Mitter, Radha Charan Pal and Surendra Nath Ray, who owned a considerable number of rental units in the city, sometimes amounting to more than a hundred, all found Phelps's proposal discriminatory and short-sighted. Landlords like them, they argued, had for generations sunk their capital into land and building, and housed the growing working-class and clerical population of the city, at a time when not the government, nor the corporation, nor the Calcutta Improvement Trust and Port Trust did anything appreciable in the sphere of housing.[110] Now the entry of Marwari traders in all commercial walks of life, including the housing market, not only

Armenians in India, from the Earliest Times to the Present Day (London: Luzac & Co., 1897), 440–45.

108 "Land Values and House Rents: Advantages of a Leasehold," *Amrita Bajar Patrika*, 2 October 1919.

109 Official and popular discourse at this time divided the capitalists from the landowners; while the capitalist was an investor, the landowner was the hoarder. This idea would continue to hold sway throughout the interwar period. See for instance "Letters to Editor: The Housing Problems," *Times of India*, 17 October 1919 and the unchanged language from eight years later in "Housing Labor in India: Colossal Problem," *Times of India*, 7 January 1927.

110 Papers Relating to the Calcutta Rent Bill, 35–32, IOR/L/PJ/6/763.

threatened them financially, but also made them victims of a colonial regulatory regime at this time of crisis within the housing market.

On 9 February 1923, nationalist Sir Surendranath Banerjea, who was the then minister for Local Self-Government, denounced and finally withdrew the bill, stating that it was a "private measure" that proposed to continue the wartime rent control until March 1924. This would not only add to the suffering of the Bengali landlords, but would also curb every incentive for building investments in the city. He added that the "facts they [the government] had gathered were meager and inconclusive and afforded no justification for a continuance of the Rent Act, except for a very limited period."[111] Disappointingly, echoing Phelps, Banerjea concluded his motion with the hope that "capitalists would invest their money in the construction of buildings, and so equalize the supply and demand in the matter of housing accommodation."[112] With the passing of a compromised bill, the debates around housing speculation came to an abrupt resolution – a resolution that neither stemmed the rising prices nor made anyone happy. In his closing words before the passing of the bill, P. C. Mitter, one of the members of the Bengal Legislative Council, said of Phelps's final argument:

> [Phelps's] speech, so far as I could follow it only amounts to this: that the landlords in the past – I am sure he has always got the landlords in his brain – have been cheating the Corporation. Hard words do not always carry conviction.[113]

The Rent Commission did not succeed in controlling the housing market through wartime regulation, nor did it succeed in ushering in preferential laws for the British and European landlords. What it managed to achieve was, first, to articulate the anxiety about the emergence of the Marwari in the lucrative housing and rental market as concerns about hoarding land, contracting the market and slowing down building activities and, second, to create a new visual template about marshes as real estate-in-waiting.

[111] "Calcutta's House Shortage: Government Continuing the Rent Act," *Straits Times*, 3 March 1923, 13.

[112] "Calcutta's House Shortage," 13. Moreover, Banerjea hoped that the two bills that were passed, the Town-Planning Bill and the Howrah Improvement Trust Bill, would accommodate the pressures on the housing market.

[113] "An Extract from an Abstract of the Proceedings of a Meeting of the Bengal Legislative Council Held on Tuesday, April 13, 1920," 64. IOR/L/PJ/6/763.

A NARRATIVE OF SQUEEZING

A small voice disrupts the confident narrative of the Rent Commission report of 1919. The marginalia attached as an addendum to the report notes that the committee was unable to arrive at a satisfactory definition of the word "profiteering":

> We may point out that under the provision of the existing law in India, increment in land value is the property of the land-owners. At the same time we would add that while unable to express an opinion on this question, many instances have been brought to our notice in which the existing dearth of accommodation has been used as a means of "squeezing" tenants unduly.[114]

The 1919 report fails to either coin a concept or develop a full-fledged definition of profiteering to bolster its numerical representation of the phenomenon. Indeed, Phelps's team deduced their argument from an extremely narrow sample size; of the 367 surveys they distributed, they received only 46 written (13 public bodies and 33 private persons) and 10 oral responses. Calcutta's population at that time was over 1.5 million. The members of the Rent Commission were unable to develop a lexical understanding of the rise in property prices and instead used the term "squeezing."

The changes in taxation on urban holding and increased migration to the city meant that investing in the housing and rental markets was on the rise in the Presidency towns of British India. Land and landed property had finally emerged as sites of investment overlaid with various kinds of accumulated and exorbitant future value claims that suddenly seemed beyond government regulation. By the second decade of the twentieth century, municipal and popular discourse began to fashion landowners as land developers, with the specific task of draining the swamps. Thus, it is no wonder that Mitra's urbanizing Calcutta is one where water bodies were fast disappearing from the physical landscape and the city's remembered histories.

[114] Rent Enquiry Committee, Municipal Miscellaneous Prog. April 1920, §7, WBSA.

Conclusion

Disappearing Coastlines

> But in time, sure enough, there was a collective setting aside of knowledge that accrues over generations through dwelling in a landscape.
>
> Amitav Ghosh, *The Great Derangement*

We easily forget. And because we forget, it is harder for us to imagine alternatives. The foregoing chapters tried to remember that which has been forgotten. Today, political analysts, world leaders and climate scientists rightfully worry that the rising waters are erasing the material remains of an already devastated delta in the mouth of the Bay of Bengal. As I conclude this book, one of the worst flooding incidents is unfolding in Bangladesh, India and Nepal. Water is already overwhelming our cities, ports, lands, designs and ways of living. We continue to look for more concrete and resilience – to fortify our lands and ourselves. We continue the processes of propertizing, pushing water out of lands, keeping them dry and looking for short-term fixes. These responses also have a history.

An alien and intrusive structure of state with its own understandings of nature, river, soil and deltaic ecology has a capacity to disturb, unsettle and unequivocally transform landscapes, water bodies and the people living there. The growth of the settlement of Calcutta in the swamps of Bengal is not just a history of urban formation, but also a history of how property, and its many instantiations in law and economy, became a tool in transforming the landscape and ecology in these spaces. By looking at the construction of a city along a dying delta over a 150 years, this book documented the transformation of the unique ecology of Bengal's littoral into solid foundations for private property. In the process, it has shown how bureaucrats, surveyors, lawyers and engineers struggled to institute hard lines of separation between land and water through the nineteenth century as an important facet of producing property in marshes.

These hard lines of separation are not simply physical lines on the ground. Instead, the drawing of these lines of separation through technology and law also entails forgetting the soaking ecologies in order to embrace the dry cultures of land use. These lines are cognitive and have shaped and transformed how we view this landscape. When they first appeared in the book, the marshes dissolved early attempts at making property along the river, eating away the very foundations on which Lacam's harbor stood until 1803. The people living there prior to the arrival of Lacam and his grand design remarked on the hubris of constructing solid structures in this changeable landscape. Over the course of the nineteenth century, property became the organizing principle for viewing spaces, be they land, water or the space in between. The book traced this history by pursuing a central question: Why and how did a land market and a regime of property become mechanisms through which these landscapes were transformed and sometimes wrested from their previous ecologies and systems of ownership?

In order to address this question, this book studied the legal fictions, infrastructural projects, economic arguments for drainage and the managerial aspects of territorial conquest. This form of conquest introduced a notarial geography in the ever-shifting landscape of the city in an estuary. The book documented the multiple attempts by the British to fortify this landscape by marshaling cartographers, hydraulic engineers, lawyers, geographers and town-planners over the course of the nineteenth century. The history of managing these marshes in and around Calcutta also revealed how petty legal cases, bureaucratic debates and techno-scientific projects undertaken by the engineers of the Bengal Army became entangled in the making of the city of Calcutta.

The fluid landscape shaped law, infrastructure and the land market. In the process land, river, marshes and property appeared as historically contingent concepts that were shaped through engineering technologies and transformed through various legal nomenclatures. In a place like Calcutta, as in many other deltaic cities such as New Orleans, Cairo, Hanoi and San Francisco, it is hard to say where the water ends and urban property begins. While we are beginning to understand the impacts of environmental engineering and human design upon such landscapes, this book attempted to look at how law worked in tandem with infrastructural projects in Calcutta to transform the space. It calls attention to our ways of viewing such ecological formations as char, sandbar and marshes as spaces waiting to be made solid. Colonial engineers and their draining technologies played only a small part in this larger legal drama.

In charting this drama, this book showed how surveys, pattahs, maps, property deeds, leases, and reports cannot simply be read as disparate acts of documentation performed by the bureaucratic state, but instead existed along a continuum that attempted to represent space, with one history and one narrative, a space uniform and flat. This was a space whose possibility of existence was *only* as property, property that could be speculated on and exchanged in a regulated market paternalized by the state. Through the materialization of paper documents, these marshy seasonal sediments of the Hooghly River entered into a history of political modernity as an economic entity that could be numerically represented through measurement and value. The sedimentation of this history is also a history of the separation of the river from the city and a history of that space between the river and the city, between water and land and how that in-between space was converted into urban property with far-reaching consequences. The colonial propertizing tendencies created a new market logic where the economic and the spatial aligned to produce land where none existed. This new spatio-economic structure, burgeoning networks of credit and an emerging politics of authority restructured the topography of the city, transforming swamps into real-estate-in-waiting.

Hence today the urban landscape at the edge of the Bay of Bengal is completely cut off from its watery history. The dynamic landscape, which was critical to the founding of the second capital of the British Empire, can perhaps best be understood as a "collage of disconnected fluvial events that were regulated principally as practices of political economy."[1] Consequently, by the early twentieth century, the intractable marshes became managed and therefore forgotten sites of a growing city.

However, such projects of management could not always bound land as property, for property continued to exceed its bound beyond the paper document. Thus, the book also tried to write that excess into the history of urban environmental transformation.[2] Property exceeded its documentary bound to reshape how we think and use land. The impacts of propertizing the swamps to build Calcutta still manifest themselves today. The

[1] Rohan D'Souza, "Seeing like a River: The Bengal Presidency's Hydraulic Transition," in *Science and Society in India 1750–2000*, ed. Arun Bandopadhyay (New Delhi: Manohar, 2010), 169–82, 174.

[2] William Cronon, *Nature's Metropolis: Chicago and the Great West* (New York, NY: Norton, 1992).

current landscape of this tidal delta comprising of not just Calcutta but cities like Dhaka, Barisal, Rajshahi and the Sandeep Islands seems remotely connected to the delta, if at all. The forgotten marshes, the floodplains and the wetlands appear as shrinking coastlines and disappearing land masses in the Bay of Bengal area. The only way water figures in these spaces now is as the disaster of floods, storms, rising seas, disappearing coastline and waterlogging. The implications of the dry culture of urban property also manifest themselves economically, ecologically and legally. The economic manifestations are felt as vast amounts of farmland are salinated by the encroaching seas. Devastating cyclones and storm surges are reshaping the coastline resulting in a slow ecological disaster. We have yet to fully grapple with the legal implications of the displacement of countless numbers of people following land subsidence that is occurring at an unimaginable speed in this area. This ecology is the colonial present. The landscape of this fluid space is marked by the capitalization of wetlands, bringing them into the infrastructures of a global financial network.[3] These moves mark a return within the postcolonial urban order to a regressive version of the political structures of colonial capital, where development and accumulation have become spatially conflated.

The wetlands that flank present-day Kolkata are the playground for a developmentalist agenda, accumulative desires and architectural design. From the airport of Kolkata to the shining new metropolis for the emergent middle class, various market forces are quickly gobbling up land, the effects of which include not just the dispossession of farmers and fishermen who lived there, but also the ecological transformation that is slowly sending parts of Bengal and Bangladesh under water, where shoreline erosion is outstripping the rate of accretion.[4] The construction boom fueled by the propertizing tendencies and the shrinking coastlines does not appear as a related event to the builders, policymakers and designers who are reshaping the landscape. The dizzying verticality of our 'atmospheric' homes in the clouds continue to suppress the soaking ecologies upon which they are built.

[3] The current West Bengal government is going ahead with a controversial plan to build a bridge on the East Kolkata Wetlands, redrawing its boundary to facilitate that and flouting the international stipulation of the 2002 Ramsar convention, which listed the East Kolkata Wetlands a "Ramsar Site," designating it as a protected area because of its "significance in terms of ecology, botany, zoology, limnology and hydrology." For more on Ramsar sites, see "Designating Ramsar Sites," www.ramsar.org/sites-countries/designating-ramsar-sites (accessed on January 12, 2018).

[4] Mahabub Sarwar and Collin D. Woodroffe, "Rates of Shoreline Change along the Coast of Bangladesh," *Journal of Coastal Conservation* 17, no. 3 (2013): 515–26.

Glossary

Bigha Unit of measurement, about one-third of an acre
Burdwan/Bardhaman An important precolonial town in Bengal that rose to prominence in the eighteenth century
Char Alluvial accretion; land formation of fluvial action in deltaic rivers found in the Lower Ganga Delta area of Bengal
Chitta Land measurement
Debuttar According to Hindu scriptures, property donated in the name of god for public good
Diara Alluvial accretion or formation
Factory town An early British settlement in India in the form of a mud factory with ballast, containing trading-related warehouse and offices
Ghat/Ghaut Riverbank
Gonda Monetary measurement in colonial Bengal
Itimandar Revenue collector
Izardar Lease holder
Jagirdar Mughal title for landlord
Kottah Unit of measurement, smaller than bigha
Khal Manmade canals
Khamar Waste (land)
Pattah/Pottah Land record, originating in Mughal period and incorporated into British administration
Presidency town When the Factory towns expanded they became Presidency towns through special directives of the London East India Company Office
Pucka/pucca House or construction with solid foundation
Rajah/Raja King or landlord

Ryot/Raiyat Persian word for peasant or cultivator
Subhedar Mughal title for the governor of a province
Wakf/Waqf Muslim charitable trust, often used to denote property
Zamindar Persian for landlord
Zamindari Authority and jurisdiction of a zamindar

Bibliography

ARCHIVES

India Office Library Records, London (IOR)

Administrative Report of the Calcutta Municipality 1867–1920
Administrative Report of the Commissioner of Port of Calcutta 1880–1910
Administrative Report of the Municipality of the Suburbs of Calcutta 1869–1880
Annual Report of the Operations of Calcutta Improvement Trust 1911–1923
Bengal Irrigation Department Manuals 1892–1926
Benjamin Lacam Papers
The Calcutta Rent Law 1920
F. Barlow Private Papers
19th-Century House of Commons Sessional Papers, Hansard Collection
Private Papers of Orme
Report of the Commissioners for the Improvement of the Town of Calcutta, 1848–1862
Surveys of Hooghly River 1834–37
Waterways and Irrigation: Rivers 1837–1916

Kolkata Port Trust

Mutty Lal Seal's Bathing Ghat, Secretary's Department, Calcutta Port Trust

National Archives of India, New Delhi (NAI)

Home Judicial Proceedings 1836–1868
Home Public Proceedings 1880–1920

Newspapers

Amrita Bazar Patrika
Basantak
The Statesman

West Bengal State Archives (WBSA)

Calcutta Improvement Trust Report 1911–1923
Government of Bengal Proceedings, Home Political [Confidential] 1900–1925
Government of Bengal Proceedings, Judicial [Judicial] and [Criminal] 1805–1880
Government of Bengal Proceedings, Public Works Department 1900–1914

PUBLISHED PRIMARY SOURCES

Documents and Reports

Anon. *Thoughts on a Late Proposal for Piloting Ships at All Seasons through Channel Creek, Instead of the Old Passage by the Braces, into Hoogly River*. London: J. Stockdale, 1782.

Atlas of the City of Calcutta. Kolkata: NATMO, 1990.

Bagchi, Prabodh C., ed. The Second City of the Empire. In *Twenty-Fifth Session of the Indian Science Congress Association*. Calcutta: Locale Secretaries of the Congress, 1938.

Ballard, John Archibald. "Remarks on a Gold Currency for India." *Bristol Selected Pamphlets*. Bombay: Thacker, Vinning & Co., 1868.

Beverly, H. *Report of the Census of the Town Calcutta, Taken on April 6th 1876*. Calcutta: Bengal Secretariat Press, 1872.

Bompas, Cecil H. "The Work of the Calcutta Improvement Trust." *Journal of Royal Society of Arts* 75, no. 7 (1927): 199–219.

Carey, William H. *Good Old Days of Honourable John Company: Being Curious Reminiscences during the Rule of the East India Company from 1600 to 1858*. Calcutta: Quins Book Company, 1882.

Chakrabarty, Sharmishta, and Bidisha De, eds. *Select Documents on Calcutta 1800–1900*. Kolkata: Directorate of State Archives, 2012.

Dacosta, John. *The Financial Situation in India, as Exposed in the Budget Statement for 1886–87*. London: P. S. King and Son, Parliamentary Agency, 1887.

Dalrymple, Alexander. *Collection of Nautical Papers Concerning the Bay of Bengal Published at the Charge of the East India Company from the Mss. by Darlymple 1784*. London: G. Biggs: British Library, 1785.

Douglas, William. "The Currency of India, with a Letter on Bi-Metalism." *Bristol Selected Pamphlets*. Glasgow: James Maclehouse, 1881.

Field, Charles D., ed. *The Regulations of the Bengal Code with a Chronological Table of Repeals and Amendments, an Introduction, Notes and an Index*. Calcutta: Thacker, Spink & Co., 1875.

First Report on the Drainage and Sewerage of Calcutta. Calcutta: Military Orphan Press, 1858.

Hirst, Frederick Christian. *Report on the Nadia Rivers 1915*. Calcutta: Bengal Government Press, 1916.

Long, James. *Selections from Unpublished Records of Government for the Years 1748 to 1767 Inclusive, Relating Mainly to the Social Condition of Bengal; with a Map of Calcutta in 1784*, vol. 1. Calcutta: Superintendent of Government Printing, 1869.

Martin, James Ranald. *Notes on the Medical Topography; Report of the Committee Appointed by the Right Honourable the Governor of Bengal for the Establishment of a Fever Hospital and for Inquiry into Local Management and Taxation in Calcutta*. London: G. H. Huttmann, 1837.

Martin, Montgomery, ed. *The Dispatches and Minutes, and Correspondence of the Marquis Wellesley, K.G. during His Administration in India*, vol. 3. London: W. H. Allen and Co., 1836.

Rainey, James H. *A Historical and Topographical Sketch of Calcutta in Four Parts*. Calcutta: Englishman Press, 1876.

Records of Fort St. George: Letters to Fort St. George for 1688. Public Sundries No. 3. Madras: Superintended Government Press, 1915.

Report of the Calcutta Building Commission. Calcutta: Government of Bengal, 1897.

Report of the Calcutta Housing and Communication Committee, 1923. Calcutta: Bengal Secretariat Press, 1924.

Report of the Committee Appointed to Enquire into the Land Value and Rents in Calcutta 1919. Calcutta: Bengal Secretariat Press, 1920.

Report of the Corporation of Calcutta Investigation Commission. Calcutta: Bengal Secretariat Press, 1950.

Richards, E. P. *Report Prepared by Request of the CIT on the Condition, Improvement and Town Planning of the City of Calcutta and Contiguous Areas*. Ware: Printed by Jennings & Bewley, 1914.

Ritchie, John. *An Hydrographical Journal of a Cursory Survey of the Coasts and Islands in the Bay of Bengal*. London: Ballantine & Law, 1784.

Saletore, Bhaskar A., ed. *Fort William House India Correspondence: 1782–1785*, vol. 9. Delhi: National Archives of India, 1959.

Seton-Karr, Walter S. *Selections from the Calcutta Gazette Showing the Political and Social Conditions of the English in India Sixty Years Ago*, vol. 5. Calcutta: Office of Superintendent of Government Printing, 1868.

Speech by the Hon'ble Surendra Nath Banerjea Introducing the Bill to Amend the Calcutta Municipal Act, 1925. Calcutta: Bengal Secretariat Press, 1925.

Smith, David B. *Report on the Drainage and Conservancy of Calcutta*. Calcutta: Bengal Secretariat Press, 1869.

T. R. "On the Manufacture of the Sylhet Lime." In *Gleanings in Science*, February 14, 1830. Calcutta: Baptist Mission Press, 1830, 61–63.

Wilson, Charles Robert. *Indian Record Series: Old Fort William in Bengal – A Selection of Official Documents Dealing with Its History*. London: John Murray, 1906.

Published Laws and Acts

Bengal Code of 1824, Regulation I
Bengal Code of 1825, Regulation XI
The Calcutta Improvement Act, 1911 of Bengal Act V 1911
Land Acquisition Act of 1894

Memoirs

Blenchynden, Kathleen. *Calcutta: Past and Present.* Calcutta: Thacker, Spink & Co., 1905.

Busteed, Henry E. *Echoes from Old Calcutta: Being Chiefly Reminiscences of the Days of Warren Hastings, Francies, and Impey.* Calcutta: Thacker, Spink & Co., 1905.

Cotton, Harry Evan Auguste. *Calcutta: Old and New: A Historical and Descriptive Handbook of the City.* 1907 reprint. Calcutta: General Publishers, 1980.

Deb, Binay Krishna. *The Early History and Growth of Calcutta.* Calcutta: Romesh Chandra Ghose, 1905.

Goode, S. W. *Municipal Calcutta: Its Institutions in Their Origin and Growth.* Edinburgh: T. and A. Constable, 1916.

Graham, Maria. *Journal of Residence in India.* Edinburgh: Archibald Constable and Co., 1812.

Kipling, Rudyard. *The Phantom Rickshaw, City of Dreadful Night and Other Tales.* New York, NY: The Lovell Company, 1890.

Sherwood, Mary. *Life of Mrs. Sherwood with Extracts from Mr. Sherwood's Journal during His Imprisonment in France and Residence in India.* London: Darton, 1857.

Toynbee, George. *A Sketch of the Administration of Hooghly District, 1795–1845*, 2nd edition. Calcutta: Ganesh Nandy, 2006 [1888].

Wheeler, Talboys J. *Madras in the Olden Time,* vol. 1: *1639–1702.* Madras: J. Higginbotham, 1861.

Wilson, Charles Robert. *Early Annals of the English in Bengal Being the Bengal Public Consultations for the First Half of the Eighteenth Century.* Calcutta: Thacker, Spink & Co., 1895.

SECONDARY SOURCES

Abrar, C. R., and S. Nurullah Azad. *Coping with Displacement: Riverbank Erosion in Northwest Bangladesh.* Dhaka: Rangpur Dinajpur Rural Service and North Bengal Institute, 2003.

Acciavatti, Anthony. *Ganges Water Machine: Designing New India's Ancient River.* Cambridge, MA: Harvard Design School, 2015.

Ackroyd, Peter. *Thames: The Sacred River.* London: Chatto & Windus, 2007.

Agha, Sameetah, and Elizabeth Kolsky. *Fringes of Empire: Peoples, Places, and Spaces in Colonial India.* New Delhi: Oxford University Press, 2009.

Ahuja, Ravi. *Pathways of Empire, Circulation, Public Works and Social Space in Colonial Orissa c. 1780 to 1914.* Delhi: Orient Black Swan, 2009.

Akhtar, Shirin. *The Role of Zamindars in Bengal, 1707–1772*. Dhaka: Asiatic Society of Bangladesh, 1982.

Alam, Mahmood, M. Mustafa Alam, Joseph R. Curray, M. Lutfar Rahman Chowdhury, and M. Royhan Gani. "An Overview of the Sedimentary Geology of the Bengal Basin in Relation to the Regional Tectonic Framework and Basin-Fill History." *Sedimentary Geology* 155, no. 3 (2003): 179–208.

Amin, Shahid. *Sugarcane and Sugar in Gorakhpur: An Inquiry into Peasant Production for Capitalist Enterprise in Colonial India*. Delhi: Oxford University Press, 1984.

Event, Metaphor, History: Chauri Chaura 1922–1992. Delhi: Oxford University Press, 1995.

Anand, Nikhil, and Anne Rademacher. "Housing in the Urban Age: Inequality and Aspiration in Mumbai." *Antipode* 43, no. 5 (2011): 1748–72.

Anderson, Katherine. *Predicting the Weather: Victorians and the Science of Meteorology*. Chicago, IL: University of Chicago Press, 2005.

Anderson, Michael R. "Classification and Coercions: Themes in South Asian Legal Studies in the 1980s," *South Asia Research* 10, no. 2 (1990): 158–77.

"The Conquest of Smoke: Legislation and Pollution in Colonial Calcutta." In *Nature, Culture and Imperials: Essays on the Environmental History of South Asia*, ed. David Arnold and Ramachandra Guha. Delhi: Oxford University Press, 1995.

"Public Nuisance and Private Purpose: Policed Environments in British India, 1860–1947," *SOAS School of Law Legal Studies Research Paper Series*, Research Paper No. 5 (2011).

Anderson, Michael R., and Sumit Guha, eds. *Changing Concepts of Rights and Justice in South Asia*. New Delhi: Oxford University Press, 2000.

Anzaldua, Gloria. *Boderlands/La Frontera: The New Mestiza*. San Francisco, CA: Aunt Lute Books, 1987.

Appadurai, Arjun. "Kings, Sects, and Temples in South India 1350–1700 A.D." *Indian Economic and Social History Review* 14, no. 1 (January–March 1977): 47–73.

"Street Culture." *India Magazine* 8, no. 1 (December 1987): 12–22.

Modernity at Large: Cultural Dimensions of Globalization. Minneapolis, MN: Minnesota University Press, 1996.

"Deep Democracy: Urban Governmentality and the Horizon of Politics." *Public Culture* 14, no. 1 (2002): 21–47.

Future as a Cultural Fact: Essays on the Global Condition. New York, NY: Verso, 2013.

Armitage, David. *The Ideological Origins of the British Empire*. Cambridge: Cambridge University Press, 2000.

Arnold, David. "Looting, Grain Riots and Government Policy in South India 1918." *Past & Present*, no. 84 (1979): 111–45.

"Perspectives of Indian Plague, 1896–1900." In *Subaltern Studies V*, ed. Ranajit Guha. New Delhi: Oxford University Press, 1987.

Colonizing the Body: State Medicine and Epidemic Disease in Nineteenth-Century India. Berkeley, CA: University of California Press, 1993.

The Tropics and the Traveling Gaze: India, Landscape, and Science, 1800–1856. Seattle, WA: University of Washington Press, 2006.

Arrighi, Giovanni. *The Long Twentieth Century: Money, Power, and the Origins of Our Times*. London: Verso, 2010.

Ascoli, Frank David. *Early Revenue History of Bengal and the Fifth Report, 1812*. Oxford: Clarendon Press, 1917.

Bagchi, Amiya. *Private Investment in India 1900–1939*. Cambridge: Cambridge University Press, 1972.

Colonialism and Indian Economy. New Delhi: Oxford University Press, 2010.

"Indian Economy and Society during World War One." *Social Scientist* 42, nos. 7/8 (2014): 5–27.

Bailey, Peter. *Leisure and Class in Victorian England: Rational Recreation and the Contest for Control, 1830–1885*. London: Methuen, 1987.

Baka, Jennifer. "The Political Construction of Wasteland: Governmentality, Land Acquisition and Social Inequality in South India." *Development and Change* 44, no. 2 (2013): 409–28.

Baker, Alan R. H. *Home and Colonial: Essays on Landscape, Ireland, Environment and Empire in Celebration of Robin Butlin's Contribution to Historical Geography*. London: Historical Geography Research Group, 2004.

Baker, Christopher J., and David Washbrook. *South India: Political Institutions and Political Change 1880–1940*. Macmillan: New Delhi, 1975.

Balachandran, Gopalan. "Britain's Liquidity Crisis and India, 1919–20." *Economic History Review* 46, no. 3 (August 1993): 575–91.

"Power and Markets in Global Finance: The Gold Standard, 1890–1926." *Journal of Global History* 3, no. 3 (2008): 313–35.

Banaji, Jairus, and Tirthankar Roy. "Markets in History: Concepts and Outcomes of Commercialisation in South Asia." *Indian Economic and Social History Review* 36, no. 3 (Special Issue on Markets in History) (1999): 271–73.

Bandyopadhyaya, Bhabanicharan. *Kalikata Kamalalay*. Calcutta: Firma KLM, 1990.

Banerjee, Anuradha. *Environment, Population, and Human Settlements of Sundarban Delta*. New Delhi: Concept Publishing Co., 1998.

Banerjee, Sumanta. *The Parlour and the Streets: Elite and Popular Culture in Nineteenth-Century Calcutta*. Calcutta: Seagull Publishers, 1998.

Crime and Urbanization: Calcutta in the Nineteenth Century. New Delhi: Tulika Books, 2006.

Banga, Indu. *The City in Indian History: Urban Demography, Society, and Politics*. New Delhi: South Asia Publications, 1991.

Ports and Their Hinterlands in India (1700–1950). New Delhi: Manohar Publications, 1992.

Banner, Stuart. *American Property: A History of How, Why and What We Own*. Cambridge, MA: Harvard University Press, 2011.

Barber, William J. *British Economic Thought and India: 1600–1858*. Oxford: Oxford University Press, 1975.

Barmaṇa, Pradip K. *Bagher Rajatve: Sundarabana Bishayaka Galpakshaalana*. Mayahuria: Samakaler Gyanakatha Prakasana, 2012.

Barnes, David S. *The Great Stink of Paris and the Nineteenth-Century Struggle against Filth and Germs*. Baltimore, MD: Johns Hopkins University Press, 2006.
Basu, Dilip K., ed. *The Rise and Growth of Colonial Port Cities in Asia*. Berkeley, CA: University of California Press, 1985.
Bayly, Christopher A. *The Local Roots of Indian Politics: Allahabad, 1880–1920*. Oxford: Oxford University Press, 1975.
Rulers, Townsmen, and Bazaars: North Indian Society in the Age of British Expansion, 1770–1870. Cambridge: Cambridge University Press, 1983.
"The British Military-Fiscal State and Indigenous Resistance: India 1720–1820." In *An Imperial State at War: Britain from 1689 to 1815*, ed. Lawrence Stone, 322–54. London: Routledge, 1994.
Bear, Laura. *Navigating Austerity: Currents of Debt along a South Asian River*. Stanford, CA: Stanford University Press, 2015.
Benton, Lauren. *Law and Colonial Cultures: Legal Regimes in World History, 1400–1900*. Cambridge: Cambridge University Press, 2002.
Search for Sovereignty: Law and Geography in European Empires, 1400–1900. Cambridge: Cambridge University Press, 2010.
Benton, Lauren, and Benjamin Straumann. "Acquiring Empire by Law: From Roman Doctrine to Early Modern European Practice." *Law and History Review* 28, no. 1 (2010): 1–38.
Beverley, Eric Lewis. "Colonial Urbanism and South Asian Cities." *Social History* 36, no. 4 (November 2011): 482–97.
"Property, Authority and Personal Law: *Waqf* in Colonial South Asia." *South Asia Research* 31, no. 2 (2011): 155–82.
Bhadra, Gautam. *Nyada Battalai Jai Kaw'bar?* Kolkata: Chhatim, 2011.
Bhattacharya, Neeladri. "Remaking Custom: The Discourse and Practice of Colonial Codification." In *Tradition, Dissent and Ideology: Essays in Honour of Romila Thapar*, ed. R. Champakalakshmi and S. Gopal, 20–54. Delhi: Oxford University Press, 1996.
Bhattacharyya, Debjani. "Geography's Myth: The Many Origins of Calcutta." In *Unarchived Histories: The Mad and the Trifling in the Colonial and Post-colonial World*, ed. Gyanendra Pandey, 144–58. New York, NY: Routledge, 2013.
"Fictions of Possession: Land, Property and Capital in Colonial Calcutta c. 1820 to c. 1920." PhD dissertation, Emory University, 2014.
"The History of Eminent Domain in British Colonial Thought and Legal Practice in South Asia." *Economic and Political Weekly* 50, no. 50 (December 2015): 45–53.
Bhattacharyya, Rajesh, and Kalyan Sanyal. "Bypassing the Squalor: New Towns, Immaterial Labour and Exclusion in Post-Colonial Urbanisation." *Economic and Political Weekly* 45, no. 31 (July 2011): 41–48.
Bhattacharyya-Panda, Nandini. *Appropriation and Invention of Tradition: The East India Company and Hindu Law in Early Colonial Bengal*. Delhi: Oxford University Press, 2008.
Biggs, David A. *Quagmire: Nation-Building and Nature in the Mekong Delta*. Seattle, WA: University of Washington Press, 2010.

Bijker, Wiebe E. "Dikes and Dams, Thick with Politics." *Isis* 98, no. 1 (2007): 109–23.

Birla, Ritu. *Stages of Capital: Law, Culture, and Market Governance in Late Colonial India*. Durham, NC: Duke University Press, 2009.

"Law as Economy: Convention, Corporation, Currency." *University of California, Irvine Law Review* 1, no. 3 (2011): 1015–37.

"Maine (and Weber) against the Grain: Towards a Postcolonial Genealogy of the Corporate Person." *Journal of Law and Society* 40, no. 1 (March 2013): 92–114.

"Speculation Illicit and Complicit: Contract, Uncertainty, and Governmentality." *Comparative Studies in South Asia, Africa and the Middle East* 35, no. 3 (2015): 392–407.

Bissell, William C. "Conservation and the Colonial Past: Urban Planning, Space and Power in Zanzibar." In *Africa's Urban Past*, ed. David Anderson and R. Rathbone, 246–61. Oxford: Oxford University Press, 2000.

Biswas, Oneil. *Calcutta and Calcuttans: From Dihi to Megalopolis*. Calcutta: Firma KLM, 1992.

Blomley, Nicholas. "Making of Private Property: Enclosure, Common Right and the Work of Hedges." *Rural History* 18, no. 1 (2007): 1–21.

Bolts, William. *Considerations on India Affairs, Particularly Respecting the Present State of Bengal and Its Dependencies*. London: J. Almon, 1772.

Booker, Matthew Morse. *Down by the Bay: San Francisco's History between the Tides*. Berkeley, CA: University of California Press, 2013.

Bose, Ajitkumar. *Kalikatar Raajpath: Samaje o Sanskritite*. Calcutta: Ananda Publishers, 1996.

Bose, Nirmal Kumar. "Calcutta: A Premature Metropolis." *Scientific American* 213, no. 3 (1965), 90–105.

Bose, Sanat Kumar. "Baul Songs of Bengal." *Folkmusic and Folklore: An Anthology*, vol. 1, ed. Hemango Biswas et al., 164–76. Calcutta: Folkmusic and Folklore Research Institute, 1967.

Bose, Sugata. *Peasant Labour and Colonial Capital: Rural Bengal since 1770*. Cambridge: Cambridge University Press, 1993.

Bourdieu, Pierre. *Outline of a Theory of Practice*. Cambridge: Cambridge University Press, 1977.

Bowen, Huw V. *Business of Empire: The East India Company and Imperial Britain*. Cambridge: Cambridge University Press, 2006.

Bowen, Huw V., John J. McAleer and Robert J. Blyth, eds. *Monsoon Traders: The Maritime World of the East Indian Company*. London: Scala, 2011.

Brenner, Joel Franklin. "Nuisance Law and the Industrial Revolution." *Journal of Legal Studies* 3, no. 2 (1974): 403–33.

Brewer, John. *The Sinews of Power: War, Money, and the English State, 1688–1783*. New York, NY: Alfred A. Knopf, 1989.

Brimnes, Neil. "Beyond Colonial Law: Indigenous Litigation and the Contestation of Property in the Mayor's Court in Late Eighteenth-Century Madras." *Modern Asian Studies* 37, no. 3 (2003): 513–50.

Broomfield, J. H. *Elite Conflict in Plural Society: Twentieth-Century Bengal*. Berkeley, CA: University of California Press, 1968.

Burns, Arthur R. "The Indian Currency Report of December, 1919 – And After." *Economica*, no. 3 (October 1921): 269–82.
Bysack, Gaur Das. "Kalighat and Calcutta." *Calcutta Review* 183 (January 1891): 305–27.
Candiani, Vera S. *Dreaming of Dry Land: Environmental Transformation in Colonial Mexico City*. Stanford, CA: Stanford University Press, 2014.
Carswell, John. *The South Sea Bubble*. London: Cresset Press, 1960.
Carter, Paul. *The Road to Botany Bay: An Exploration of Landscape and History*. New York, NY: Faber, 1988.
Dark Writing: Geography, Performance and Design. Honolulu, HI: University of Hawai'i Press, 2008.
Casey, Edward. *Fate of Place: A Philosophical History*. Berkeley, CA: University of California Press, 1998.
Castree, Noel. "Marxism and the Production of Nature." *Capital & Class* 24, no. 3 (October 1, 2000): 5–36.
Chakrabarty, Dipesh. "The Colonial Context of the Bengal Renaissance: A Note on Early Railway Thinking in Bengal." *Indian Economic and Social History Review* 11, no. 1 (1974): 92–106.
"Open Space/Public Space: Garbage, Modernity and India." *South Asia* 14, no. 1 (1991): 15–31.
Rethinking Working-Class History. Princeton, NJ: Princeton University Press, 2000.
Habitations of Modernity: Essays in the Wake of Subaltern Studies. Chicago, IL: University of Chicago Press, 2002.
Provincializing Europe: Postcolonial Thought and Historical Difference. Princeton, NJ: Princeton University Press, 2007.
Chakravorty, Sanjoy. *The Price of Land: Acquisition, Conflict, Consequence*. Oxford: Oxford University Press, 2013.
Chandravarkar, Rajnarayan. *The Origins of Industrial Capitalism in India: Business Strategies and the Working Classes in Bombay, 1900–1940*. Cambridge: Cambridge University Press, 1994.
Imperial Power and Popular Politics: Class, Resistance and State in India c. 1850–1950. Cambridge: Cambridge University Press. 1998.
Chapman, Barbara. *The Colonial Eye: A Topographical and Artistic Record of the Life and Landscape of Western Australia, 1798–1914*. Perth: Art Gallery of Western Australia, 1979.
Chatterjee, Kumkum. "Goddess Encounters: Mughals, Monsters and the Goddess in Bengal." *Modern Asian Studies* 47, no. 5 (2013): 1435–87.
Chatterjee, Partha. *Bengal 1920–1947: The Land Question*. Kolkata: K. P. Bagchi, 1984.
The Nation and Its Fragments: Colonial and Postcolonial Histories. Princeton, NJ: University of Princeton Press, 1994.
Present History of West Bengal. Delhi: Oxford University Press, 1997.
A Princely Imposter? The Strange and Universal History of the Kumar of Bhawal. Princeton, NJ: Princeton University Press, 2002.
The Politics of the Governed: Reflections on Popular Politics in Most of the World. New York, NY: Columbia University Press, 2004.

The Black Hole of Empire: History of a Global Practice of Power. Princeton, NJ: Princeton University Press, 2012.

Chatterjee Sarkar, Sutapa. "Bengal's Southern Frontier, 1757 to 1948." *Studies in History* 28, no. 1(2012): 69–97.

Chattopadhyay, K. P., and H. K. Chaturvedi. "How Jute Workers Live." *Science and Culture* 12, no. 8 (February 1947): 367–79.

Chattopadhyay, Swati. *Representing Calcutta: Modernity, Nationalism, and the Colonial Uncanny*. London: Routledge, 2006.

Unlearning the City: Infrastructure in a New Optical Field. Minneapolis, MN: University of Minnesota Press, 2012.

Chaudhuri, Kalyan N. *The English East India Company: The Study of an Early Joint-Stock Company 1600–1640*. New York, NY: Augustus M. Kelley, 1965.

"Some Reflections on the Town and Country in Mughal India." *Modern Asian Studies* 12, no. 1 (1978): 77–96.

Chaudhuri, Sukanta. *Calcutta: The Living City*, 2 vols. Delhi: Oxford University Press, 1993.

Childers, Joseph. "Foreign Matter, Imperial Filth." In *Filth: Dirt, Disgust, and Modern Life*, ed. William Cohen and Ryan Johnson, 201–21. Minneapolis, MN: University of Minnesota Press, 2005.

Chopra, Preeti. *A Joint Enterprise: Indian Elites and the Making of British Bombay*. Minneapolis, MN: University of Minnesota Press, 2011.

Christophe-Agnew, Jean. *Worlds Apart: The Market and the Theatre in Anglo-American Thought, 1550–1750*. Cambridge: Cambridge University Press, 1988.

Cohn, Bernard. "From Indian Status to British Contract." *Journal of Economic History* 21, no. 4 (1961): 613–28.

"Representing Authority in Victorian India." In *The Invention of Tradition*, ed. Eric Hobsbawm and Terence Ranger, 165–210. Cambridge: Cambridge University Press, 1983.

An Anthropologist among the Historians and Other Essays. Delhi: Oxford University Press, 1987.

Colonialism and Its Forms of Knowledge: The British in India. Princeton, NJ: Princeton University Press, 1996.

Conlon, Frank. "Industrialization and the Housing Problem in Bombay, 1850–1940." In *Changing South Asia: Economy and Society*, ed. Kenneth Ballhatchet and David Taylor, 153–68. London: Asia Research Service, 1984.

Conrad, Joseph. *Heart of Darkness*. New York, NY: Penguin, 2007.

Coombe, Rosemary. *The Cultural Life of Intellectual Properties: Authorship, Appropriation and the Law*. Durham, NC: Duke University Press, 1998.

Cooper, Frederick. *Struggles for the City: Migrant Labour, Capital, and the State in Urban Africa*. London: Sage, 1983.

Cosgrove, Denis, and Geoff Petts. *Water, Engineering and Landscape: Water Control and Landscape Transformation in the Modern Period*. London: Bellhaven Press, 1990.

Cowan, David, and Daniel Wincott, eds. *Exploring the "Legal" in Socio-Legal Studies*. London: Palgrave Macmillan, 2016.

Craib, Raymond. *Cartographic Mexico: A History of State Fixation and Fugitive Landscapes*. Durham, NC: Duke University Press, 2004.
Crais, Clifton. "The Vacant Land: The Mythology of British Expansion in the Eastern Cape, South Africa." *Journal of Social History* 25, no. 2 (Winter 1991): 255–75.
Cronon, William. *Natures Metropolis: Chicago and the Great West*. New York, NY: Norton, 1992.
Curley, David. *Poetry and History: Bengali Mangal-Kabya and Social Change in Precolonial Bengal*. Delhi: Chronicle Books, 2008.
da Cunha, Dilip. *Invention of the River: Alexander's Eye and Ganga's Descent*. Philadelphia, PA: University of Pennsylvania Press, 2018.
Das, Chandan Surabhi, and Sunando Bandyopadhyay. *Sharing Space: Human-Animal Conflicts in Indian Sundarban*. Kolkata: Progressive Publishers, 2012.
Dasgupta, Ashin. *Indian Merchants and the Decline of Surat c. 1700–1750*. Wiesbaden: Franz Steiner Verlag, 1979.
Dasgupta, Keya. "Genesis of a Neighbourhood: The Mapping of Bhabanipur." Occasional Paper 175, Centre for Studies in Social Sciences, Calcutta (March 2003).
Mapping Calcutta: The Collection of Maps at the Visual Archives of the Centre for Studies in Social Sciences. Calcutta: CSSS, 2009.
Daston, Lorraine. "Cloud Physiognomy." *Representations* 135, no. 1 (2016): 45–71.
Daston, Lorraine, and Katherine Park. *Wonders and the Order of Nature 1150–1750*. New York, NY: Zone Books, 1988.
Daston, Lorraine, and Peter Galison. "The Image of Objectivity." *Representations* 40, Special Issue: Seeing Sciences (Autumn 1992): 81–128.
Objectivity. New York, NY: Zone Books, 2007.
Datta, Partho. *Planning the City: Urbanization and Reform in Calcutta c. 1800–c. 1940*. Delhi: Tulika Books, 2012.
Davis, Mike. *Planet of Slums*. London: Verso, 2006.
De (Dnere), Shankar. *Samajik O Rajnaitik Prekhapete Jele Kaibarta: Adi Kolkatar Jelepara, Jelepara-r Swang*. Kolkata: Offbeat Publishing, 2006.
Deacon, Harriet. "Racial Segregation and Medical Discourse in Nineteenth-Century Cape Town." *Journal of Southern African Studies* 22, no. 2 (1996): 287–308.
Dean, Carolyn. "Rock Sites/Rock's Sight: Reflections on Site Documentation." *Public Art Dialogue* 1, no. 2 (2011): 151–61.
Deb, Binay Krishna. *Early History and Growth of Calcutta*. Calcutta: Bengal Printing Company, 1905.
Dembowski, Hans. "Courts, Civil Society and Public Sphere: Environmental Litigation in Calcutta." *Economic Political Weekly* 34, nos. 1/2 (January 2–15, 1999): 49–56.
Denning, Greg. *Islands and Beaches. Discourse on a Silent Land: Marquesas, 1774–1880*. Honolulu, HI: University of Hawai'i Press, 1980.
Derrett, J. Duncan M. "The Administration of Hindu Law by the British." *Comparative Studies in Society and History* 4, no. 1 (November 1961): 10–52.
Essays in Classical and Modern Hindu Law, vol. 1: *Dharmasastra and Related Ideas*. Leiden: E. J. Brill, 1976.

Didier, Emmanuel. "Do Statistics 'Perform' the Economy?" In *Do Economists Make Markets? On the Performativity of Economics*, ed. Donald Mackenzie, Fabian Muniesa and Lucia Siu, 276–310. Princeton, NJ: Princeton University Press, 2007.

Dimock, Edward. *The Place of the Hidden Moon: Erotic Mysticism in the Vaisnava-Sahajiyā Cult of Bengal*. Chicago, IL: University of Chicago Press, 1966.

Dirks, Nicholas. "From Little King to Landlord: Property, Law and the Gift under the Madras Permanent Settlement." *Comparative Studies in Society and History* 28, no. 2 (1986): 307–33.

Doss, Lal Mohun. *The Law of Riparian Rights, Alluvion and Fishery: With Introductory Lectures on the Rights of Littoral States over the Open Sea, Territorial Waters, Bays, &c., and the Rights of the Crown and the Littoral Proprietors Respectively over the Fore-Shore of the Sea*. Calcutta: Thacker, Spink & Co., 1891.

Dossal, Mariam. *Theatre of Conflict, City of Hope: Mumbai 1660 to Present Times*. New Delhi: Oxford University Press, 2010.

D'Souza, Rohan. *Drowned and Dammed: Colonial Capitalism and Flood Control in Eastern India*. New Delhi: Oxford University Press, 2006.

Dutt, Binode Behari. *Town Planning in Ancient India*. New Delhi: New Asian Publishers, 1977.

Dutt, Romesh C. *The Economic History of India under Early British Rule: From the Rise of the British Power in 1757 to the Accession of Queen Victoria in 1837*, vol. 1. London: Kegan Paul, Trench Trubner, 1902.

Dutt, Shib Chandra. "The Two Poles of Indian Economic Thought." In *Conflicting Tendencies in Indian Economic Thought*. Calcutta: N. M. Chowdhury and Co., 1928.

Dutta, Rajat. *Society, Economy and the Market: Commercialization in Rural Bengal, c. 1760–1800*. Delhi: Manohar Publishers, 2000.

"Seeing like a River: The Bengal Presidency's Hydraulic Transition." In *Science and Society in India 1750–2000*, ed. Arun Bandopadhyay, 169–82. New Delhi: Manohar, 2010.

Eaton, Richard Maxwell. *The Rise of Islam and the Bengal Frontier, 1204–1760*. Berkeley, CA: University of California Press, 1993.

Eck, Diana. *India: A Sacred Geography*. New York, NY: Three Rivers Space, 2012.

Edney, Mathew H. *Mapping an Empire: The Geographical Construction of British India 1765–1843*. Chicago, IL: University of Chicago Press, 1997.

Edwardes, Stephen Meredyth., James M. Campbell, and Bombay (Presidency). *The Gazetteer of Bombay City and Island*, 3 vols. Bombay: Time Press, 1909.

Edwards, Paul N. *A Vast Machine: Computer Models, Climate Data and the Politics of Global Warming*. Cambridge, MA: MIT Press, 2010.

el-Aswad, El-Sayed. "Spiritual Genealogy: Sufism and Saintly Places in the Nile Delta." *International Journal of Middle Eastern Studies* 38, no. 4 (2006): 501–18.

Fergusson, James. "On Recent Changes in the Delta of the Ganges." *Quarterly Journal of the Geological Society* 19 (1863): 321–54.

Fitzmaurice, Andrew. "The Genealogy of Terra Nullius." *Australian Historical Studies* 38, no. 129 (2007): 1–15.
Forlong, James George Roche. "The Life of Calcutta as a Seaport and the Mercantile Capital of Asia." *Geographical Journal* 16, no. 2 (1900): 238–41.
Forster, Edward Morgan. *A Passage to India*. London: Edward Arnold and Co., 1924.
Fox, Richard G., ed. *Urban India: Society, Space, Image*. Durham, NC: Duke University Press, 1970.
Foucault, Michel. *Discipline and Punish: The Birth of the Prison*. London: Harmondsworth, 1977.
History of Sexuality, vol. 1: *An Introduction*. London: Vintage 1979.
"Questions on Geography: Interviewers: The Editors of the Journal *Herodote*." In *Michel Foucault: Power/Knowledge: Selected Interview and Other Writings 1972–77*. Brighton: Harvester, 1980.
"Of Other Spaces." *Diacritics* 16 (Spring 1986): 22–27.
"Governmentality." In *Essential Works of Foucault, 1954–1984*, vol. 3: *Power*, vol. 3. London: Penguin, 2001.
Friedman, John, and Robert Wulff. *The Urban Transition: Comparative Studies of Newly Industrializing Societies*. London: Edward Arnold, 1976.
Furedy, Chris. "Interest Groups and Municipal Management in Calcutta 1875–1890." *York Historical Papers* (1973): 192–211.
Galanter, Marc, and Rajeev Dhavan. *Law and Society in Modern India*. Delhi: Oxford University Press, 1989.
Gandy, Matthew. *Concrete and Clay: Reworking Nature in New York City*. Cambridge, MA: MIT Press, 2003.
"Rethinking Urban Metabolism: Water, Space and the Modern City." *City* 8, no. 3 (2004): 363–79.
The Fabric of Space: Water, Modernity, and the Urban Imagination. Cambridge, MA: MIT Press, 2014.
Getzler, Joshua. *A History of Water Rights at Common Law*. Oxford: Oxford University Press, 2004.
Ghosh, Amitav. *The Hungry Tide*. London: Harper Collins, 2004.
The Great Derangement: Climate Change and the Unthinkable. Chicago, IL: University of Chicago Press, 2016.
Ghosh, Anindita. *Power in Print: Popular Publishing and the Politics of Language and Culture in a Colonial Society*. Delhi: Oxford University Press, 2006.
Ghosh, Benoy. *Kalpeynchar Naksha*. Calcutta: Bak-Sahitya, 1961.
Town Kolkatar Kedcha. Calcutta: Bak-Sahitya, 1964.
Ghosh, Pika. "Unrolling a Narrative Scroll: Artistic Practice and Identity in Late-Nineteenth-Century Bengal." *Journal of Asian Studies* 62, no. 3 (2003): 835–71.
Gidwani, Vinay. "'Waste' and the Permanent Settlement in Bengal." *Economic and Political Weekly* 27, no. 4 (1992): PE39–PE46.
Gilmartin, David. "Scientific Empire and Imperial Science: Colonialism and Irrigation Technology in the Indus Basin." *Journal of Asian Studies* 53, no. 4 (1994): 1127–48.
"Water and Waste: Nature, Productivity and Colonialism in the Indus Basin." *Economic and Political Weekly* 38, no. 48 (November 2003): 5057–65.

Blood and Water: Indus River Basin in Modern History. Berkeley, CA: University of California Press, 2015.

Glover, William J. *Making Lahore Modern: Constructing and Imagining a Colonial City*. Minneapolis, MN: University of Minnesota Press, 2007.

Golan, Tal. *Laws of Men and Laws of Nature*. Cambridge, MA: Harvard University Press, 2009.

Goswami, Jayanta. *Samajacitre Unabingsha Satabdir Bangla Prahasan*. Kolkata: Sahityasree, 1975.

Goswami, Manu. *Producing India: From Colonial Economy to National Space*. Chicago, IL: University of Chicago Press, 2004.

Goswami, Omkar. *Industry, Trade and Peasant Society: The Jute Economy of Eastern India, 1900–1947*. Oxford: Oxford University Press, 1991.

Gould, Eliga H. "Entangled Histories, Entangled Worlds: The English-Speaking Atlantic as a Spanish Periphery." *American Historical Review* 112, no. 3 (2007): 764–86.

"Entangled Atlantic Histories: A Response from the Anglo-American Periphery." *American Historical Review* 112, no. 5 (2007): 1415–22.

Guha, Ranajit. *A Rule of Property in Bengal: An Essay on the Idea of Permanent Settlement*. Durham, NC: Duke University Press, 1996.

History at the Limits of World History. New York, NY: Columbia University Press, 2002.

Guha, Sumit. *Fissured Land: An Ecological History of India*. New Delhi: Oxford University Press, 1992.

Guha-Thakurta, Tapati. "The Compulsions of Visual Representation in Colonial India." In *Traces of India: Photography, Architecture, and the Politics of Representation, 1850–1950*, ed. Maria Antonella Pelizzari, 108–39. New Haven, CT: Yale University Press, 2003.

Gupta, Brijen. *Sirajuddaullah and the East India Company, 1756–1757*. Leiden: E. J. Brill, 1962.

Gupta, Narayani. *Delhi between Two Empires 1803–1931: Society, Government and Urban Growth*. Delhi: Oxford University Press, 1998.

Gupta, Priya S. "The Peculiar Circumstances of Eminent Domain in India." *Osgoode Hall Law Journal* 49, no. 3 (Summer 2012): 445–89.

Gupta, Raj Bahadur. *Labour and Housing in India*. Calcutta: Longman Green & Co., 1929.

Gupta, Samita. "Theory and Practice of Town Planning in Calcutta, 1817 to 1912: An Appraisal." *Indian Economic and Social History Review* 30, no. 1 (March 1922): 29–55.

Greenblatt, Stephen. *Marvellous Possessions: The Wonder of the New World*. Chicago, IL: University of Chicago Press, 1991.

Grey, Thomas. "The Disintegration of Property." In *Property*, ed. J. Ronald Pennock and John W. Chapman, 291–307. New York, NY: New York University Press, 1980.

Hahn, Chris M., ed. *Property Relations: Renewing the Anthropological Tradition*. Cambridge: Cambridge University Press, 1998.

Haines, Daniel. *Building the Empire, Building the Nation: Development, Legitimacy and Hydro-Politics in Sind, 1919–1969*. Karachi: Oxford University Press, 2013.

Hardgrove, Anne. *Community and Public Culture: The Marwaris in Calcutta, c. 1897–1997*. New York, NY: Columbia University Press, 2004.

Hardt, Michael, and Antonio Negri. *Empire*. Cambridge, MA: Harvard University Press, 2000.

Harrison, Mark. *Climates and Constitutions: Health, Race, Environment and British Imperialism in India, 1600–1850*. New Delhi: Oxford University Press, 1999.

Harvey, David. *The Urban Experience*. Baltimore, MD: Johns Hopkins University Press, 1989.

Spaces of Capital: Towards a Critical Geography. Edinburgh: Edinburgh University Press, 2001.

"The New Imperialism: Accumulation by Dispossession." *The Socialist Register. The New Imperial Challenge* 40 (2004): 63–87.

Spaces of Global Capitalism: Towards a Theory of Uneven Geographical Development. New York, NY: Verso, 2006.

Hayes, Glen A. "Boating upon the Crooked River: Cosmophysiological Soteriologies in the Vaisnava Sahajiya Tradition of Medieval Bengal." In *Shaping Bengali Worlds, Public and Private*, ed. Tony K. Stewart, 29–35. East Lansing, MI: Asian Studies Center, Michigan State University.

Haynes, Douglas E. *Rhetoric and Ritual in Colonial India: The Shaping of a Public Culture in Surat City, 1852–1928*. Berkeley, CA: University of California Press, 1991.

Hazareesingh, Sandip. *The Colonial City and the Challenge of Modernity: Urban Hegemonies and Civic Contestations in Bombay City 1900–1925*. New Delhi: Orient Black Swan 2007.

Heidegger, Martin. *The Question Concerning Technology, and Other Essays*. New York, NY: Harper and Row, 1977.

Heilbron, John. "The Laws of Nature and the Laws of the Land." October 1995, unpublished paper.

Herzog, Tamar. *Frontiers of Possession: Spain and Portugal in Europe and the Americas*. Cambridge, MA: Harvard University Press, 2015.

Hill, Christopher V. *River of Sorrow: Environment and Social Change in Riparian North India, 1770–1994*. Ann Arbor, MI: University of Michigan Press, 1997.

Hopkins, Anthony G. "Property Rights and Empire Building: Britain's Annexation of Lagos, 1861." *Journal of Economic History* 40, no. 4 (1980): 777–98.

Hosagrahar, Jyoti. *Indigenous Modernities: Negotiating Architecture and Urbanism*. London: Routledge, 2005.

Hull, Mathew S. *Government of Paper: The Materiality of Bureaucracy in Urban Pakistan*. Berkeley, CA: University of California Press, 2012.

Hunter, William. "The Territorial Aristocracy in Bengal, the Burdwan Raj." *Calcutta Review* 54, no. 108 (1872): 171–92.

Hussain, Nasser. *The Jurisprudence of Emergency: Colonialism and the Rule of Law*. Ann Arbor, MI: University of Michigan Press, 2003.

Hussin, Iza. "Circulations of Law: Cosmopolitan Elites, Global Repertoires, Local Vernaculars." *Law and History Review* 32, no. 4 (2014): 773–95.

Ingold, Alice. "To Historicize or Naturalize Nature: Hydraulic Communities and Administrative States in Nineteenth-Century Europe." *French Historical Studies* 32, no. 3 (2009): 129–53.

Iqbal, Iftekhar. *The Bengal Delta: Ecology, State and Social Change 1840–1943*. New York, NY: Palgrave Macmillan, 2010.

"First Master Plan for Dhaka City: An Environmental Exploration." *South Asian Chronicle* 3 (2013): 42–61.

Jain, Lakshmi Chand. *Indigenous Banking in India*. Bombay: Macmillan & Co., 1929.

Jalais, Annu. *Forest of Tigers: People, Politics and Environment in the Sundarbans*. New Delhi: Routledge, 2010.

Jana, Manindranath. *Sundarbaner Samaja o Samskriti*. Calcutta: Dipali Book House, 1984.

Jaoul, Nicolas. "Learning the Use of Symbolic Means: Dalits, Ambedkar Statues and the State in Uttar Pradesh." *Contributions to Indian Sociology* 40, no. 2 (2006): 175–207.

Joseph, Charles. "Topographical Survey of the River Hooghly from Bandel to Garden Reach, Exhibiting the Principal Buildings, Ghats and Temples on both Banks, Executed in the Year 1841." *Calcutta Review* 6, no. 3 (January–June 1845): 428–62.

Kale, Sunila. *Electrifying India: Regional Political Economies of Development*. Stanford, CA: Stanford University Press, 2014.

Kalia, Ravi. *Chandigarh: The Making of an Indian City*. Delhi: Oxford University Press, 1998.

Karan, Mahendranath. "Khejuri Bandar." *Masik Basumati* 2, no. 4 (1925): 275–482.

Karim, Khondkar M. "Mughal *nawaara* in Bengal." *Journal of Asiatic Society of Bengal* 14, no. 1 (1969): 51–59.

Kelman, Ari. *A River and Its City: The Nature of Landscape in New Orleans*. Berkeley, CA: University of California Press, 2003.

Kerr, Ian J. *Building the Railways of the Raj: 1850–1900*. Delhi: Oxford University Press, 1995.

Keydar, Caglar. "Globalization and Social Exclusion in Istanbul." *International Journal of Urban and Regional Research* 291 (March 2005): 124–34.

King, Anthony. *Colonial Urban Development: Culture, Social Power and Environment*. London: Routledge, 1976.

Klein, Ira. "Urban Development and Death: Bombay City, 1870–1914." *Modern Asian Studies* 20, no. 4 (1986): 725–54.

Kolb, Charles R., and Jack R. Van Lopik. "Depositional Environments of the Mississippi River Deltaic Plain." In *Deltas in Their Geologic Framework*, ed. Martha Lou Shirley. Houston, TX: Houston Geologic Society, 1969.

Kosambi, Meera, and John E. Brush. "Early European Suburbanization in the Indo-British Port Cities." In *Asian Urbanization: Problems and Processes*, ed. Frank J. Costa et al., 9–23. Berlin: Gebruder Bornträger, 1988.

"Three Colonial Port Cities in India." *Geographical Review* 78, no. 1 (1988): 32–47.

Kozlowski, Gregory. *Muslim Endowments and Society in British India*. Cambridge: Cambridge University Press, 1985.

"Muslim Women and the Control of Property in North India." In *Women and Social Reform in Modern India: A Reader*, vol. 2, ed. Sumit Sarkar and Tanika Sarkar, 326–41. Ranikhet: Permanent Black, 2007.

Krause, Franz. "Making Space along the Kemi River: A Fluvial Geography in Finnish Lapland." *Cultural Geographies* 24, no. 2 (2017): 279–94.

Kumar, Dharma, and Meghnad Desai. *The Cambridge Economic History of India*, vol. 2: *c. 1751–c. 1970*. Cambridge: Cambridge University Press, 1983.

Laha, N. N. *Subarnabanik Katha o Kriti*, vol. 3. Kolkata: Dey's Publication, 1942.

Lahiri Choudhury, Deep Kanta. *Telegraphic Imperialism: Crisis and Panic in the Indian Empire, c. 1830*. Basingstoke: Palgrave Macmillan, 2010.

Lahiri-Dutta, Kuntala, and Gopa Samanta. *Dancing with the River: People and Life on the Chars of South Asia*. New Haven, CT: Yale University Press, 2013.

Larkin, Paschal. *Property in the Eighteenth Century: With Special Reference to England and Locke*. Dublin: Cork University Press, 1930.

Latour, Bruno. *The Making of Law: An Ethnography of the Conseil D'etat*. Cambridge: Polity, 2010.

"Some Experiments in Art and Politics." e-flux Journal, 23 March 2011, www.e-flux.com/journal/some-experiments-in-art-and-politics/.

Lefebvre, Henri. *The Production of Space*. Oxford: Blackwell, 1991.

Legg, Steve. *Spaces of Colonialism: Delhi's Urban Governmentalities*. Malden, MA: Wiley Blackwell, 2007.

Lesgar, Clé. *The Rise of the Amsterdam Market and Information Exchange: Merchants, Commercial Expansion and Change in the Spatial Economy of the Low Countries, c. 1550–1630*. New York, NY: Ashgate Publishing, 2016.

Lewandowski, Susan. "Changing Form and Function in the Ceremonial and Colonial Port City in India: An Historical Analysis of Madurai and Madras." *Modern Asian Studies* 11, no. 2 (1977): 183–212.

"The Hindu Temple in South India." In *Buildings and Society: Essays on the Social Development of the Built Environment*, ed. Anthony D. King, 67–82. London: Psychology Press, 1984.

Lorea, Carola Erika. "*Learning to Swim in the River of Desire*: Bhaba Pagla's Songs in Their Performative Context." PhD thesis, Sapienza University of Rome, 2014.

Ludden, David. *Peasant History in South India*. Princeton, NJ: Princeton University Press, 1985.

"India's Developmental Regime." In *Colonialism and Culture*, ed. Nicolas Dirks, 247–87. Ann Arbor, MI: University of Michigan Press, 1992.

"Development Regimes in South Asia: History and the Development Conundrum." *Economic and Political Weekly* 40, no. 37 (September 10, 2006): 4042–51.

Lyell, Charles Sir. *Principles of Geology*. London: John Murray, 1834.

Magee, Gary Bryan, and Andrew S. Thompson. *Empire and Globalisation: Networks of People, Goods and Capital in the British World, c. 1850–1914*. Cambridge: Cambridge University Press, 2010.

Malavika, Kasturi. "'Asceticizing' Monastic Families: Ascetic Genealogies, Property Feuds and Anglo Hindu Law in Late Colonial India." *Modern Asian Studies* 43, no. 5 (2009): 1039–83.

Malhotra, Daya Krishnan. *History and Problems of Indian Currency, 1835–1949: An Introductory Study*. Simla: Minerva Book, 1949.
Mallabarman, Adwaita. *Titas Ekti Nadir Naam*. Kolkata: Punthigar, 1950.
Mantena, Karuna. *Alibis of Empire: Henry Maine and the Ends of Liberal Imperialism*. Princeton, NJ: Princeton University Press, 2010.
Marshall, Peter J. *East Indian Fortunes: The British in Bengal in the Eighteenth Century*. Oxford: Oxford University Press, 1976.
Bengal: The British Bridgehead Eastern India 1740–1828. Cambridge: Cambridge University Press, 1988.
Masselos, Jim. *Towards Nationalism: Group Affiliations and the Politics of Public Associations in Nineteenth Century Western India*. Bombay: Popular Prakashan, 1974.
Mathur, Anuradha, and Dilip da Cunha. *Mississippi Floods: Designing a Shifting Landscape*. New Haven, CT: Yale University Press, 2001.
Soak: Mumbai in an Estuary. New Delhi: Rupa Publishing, 2009.
Design in the Terrain of Water. Philadelphia, PA: University of Pennsylvania Press, 2014.
Mawani, Renisa, and Iza Hussin. "Law and Settler Colonialism, and 'the Forgotten Space' of Maritime Worlds." *Annual Review of Law and Social Science* 12 (2013): 107–31.
Mawani, Renisa, and Iza Hussin, eds. Special Issue, "The Travels of Law: Indian Ocean Itineraries." *Law and History Review* 32, no. 4 (2014).
McLane, John R. *Land and Local Kingship in Eighteenth Century Bengal*. Cambridge: Cambridge University Press, 1993.
Mehta, Markand. *The Ahmedabad Cotton Textile Industry: Genesis and Growth*. Ahmedabad: New Order, 1982.
Meller, Hellen E. *Patrick Geddes: Social Evolutionist and City Planner*. London: Routledge, 1975.
Mels, Tom. "Primitive Accumulation and the Production of Abstract Space: Nineteenth-Century Mire Reclamation on Gotland." *Antipode* 46, no. 4 (2014): 1113–33.
Menon, Ajit. "Colonial Constructions of 'Agrarian Fields' and 'Forests' in the Kolli Hills." *Indian Economic and Social History Review* 41, no. 3 (2004): 315–37.
Merry, Sally Engle. "McGill Convocation Address: Legal Pluralism in Practice." *McGill Law Journal* 59, no. 1 (2013): 1–8.
Mitchell, Timothy. *Colonising Egypt*. Berkeley, CA: University of California Press, 1991.
"Society, Economy and the State Effect." In *State/Culture: State Formation after the Cultural Turn*, ed. George Steinmetz, 76–97. Ithaca, NY: Cornell University Press, 1999.
Rule of Experts: Egypt, Techno-Politics, Modernity. Stanford, CA: University of California Press, 2002.
Mitchell, William J. T., ed. *Landscape and Power*. Chicago, IL: University of Chicago Press, 1994.
Mitra, Kissory Chand. *Mutty Lall Seal*. Calcutta: D'Rozario Printer, 1869.

Mitra, Premdendra. *Premendramitra Rachanabali*, vol. 1. Kolkata: Granthalay Pvt. Ltd., 1976.
Mitra, Radharaman. *Kalikata Darpan*. Calcutta: Riddhi, 1986.
Mitra, Sarada Charan. *The Land-Law of Bengal, Tagore Law Lectures 1895*. Calcutta: Thacker, Spink & Co., 1898.
Morris, Christopher. *The Big Muddy: An Environmental History of the Mississippi and Its Peoples from Hernando de Soto to Hurricane Katrina*. New York, NY: Oxford University Press, 2012.
Morris, David Morris. *The Emergence of an Industrial Labor Force in India: A Study of the Bombay Cotton Mills, 1854– 1947*. Berkeley, CA: University of California Press, 1965.
Mukharji, Projit Bihari. *Doctoring Traditions: Ayurveds, Small Technologies and Braided Sciences*. Chicago, IL: University of Chicago Press, 2016.
Mukherjea, Bijan K. *The Hindu Law of Religious and Charitable Trust*. Calcutta: Eastern Law House, 1970.
Mukherjee, Nilmani. *The Port of Calcutta: A Short History*. Calcutta: Commissioners of Port of Calcutta, 1968.
Mukherjee, Radhakamal. *The Changing Face of Bengal: A Study in Riverine Economy*. Calcutta: University of Calcutta, 1938.
The Indian Working Class. Bombay: Hind Kitab, 1945.
Mukherjee, Tillotama. "The Co-Ordinating State and the Economy: The Nizamat in Eighteenth-Century Bengal." *Modern Asian Studies* 43, no. 2 (2009): 389–436.
Mukhopadhyay, Harishadhan. *Kalikata shekale'r o ekale'r*. Kolkata: P. M. Bagchi, 1985.
Mukhopahyay, Tarun Kumar. *Hickey's Bengal Gazette: Contemporary Life and Events in Calcutta*. Calcutta: Subarnarekha, 1988.
Munn, Nancy D. "Gawan Kula: Spatiotemporal Control and the Symbolism of Influence." In *The Kula: New Perspectives on Massim Exchange*, ed. J. W. Leach and E. Leach. Cambridge: Cambridge University Press, 1983.
Murphey, Rhoads. "The City in the Swamps: Aspects of the Site and Early Growth of Calcutta." *The Geographical Journal* 130, no. 2(1964): 241–56.
Nadasdy, Paul. "Boundaries among Kin: Sovereignty, the Modern Treaty Process, and the Rise of Ethno-Territorial Nationalism among Yukon First Nations." *Comparative Studies in Society and History* 54, no. 3 (2012): 499–532.
Naidu, Ratna. *Old Cities, New Predicaments: A Study of Hyderabad*. Delhi: Sage Publications, 1991.
Nair, Janaki. *The Promise of the Metropolis: Bangalore's Twentieth Century*. New Delhi: Oxford University Press, 2005.
Nair, P. Thankappan. *Calcutta: The Origin of the Name*. Calcutta: Subarnarekha, 1985.
A History of Calcutta's Streets. Calcutta: Firma KLM, 1987.
Calcutta in the 19th Century: Company's Days. Calcutta: Firma KLM, 1989.
Calcutta: Tercentenary Bibliography. Calcutta: Asiatic Society, 1993.
Naqvi, Hameeda Khatoon. *Urbanisation and Urban Centres under the Great Mughals 1556–1707: An Essay in Interpretation*, vol. 1. Simla: Indian Institute of Advanced Study, 1972.

Nilsson, Sten. *The New Capitals of India, Pakistan and Bangladesh*. London: Malma, 1973.

Nixon, Rob. *Slow Violence and the Environmentalism of the Poor*. Cambridge, MA: Harvard University Press, 2011.

Ocko, Jonathan, and David Gilmartin. "State, Sovereignty and the People: A Comparison of the 'Rule of Law' in China and India." *Journal of Asian Studies* 68, no. 1 (2009): 55–100.

Oldenburg, Veena Talwar. *The Making of Colonial Lucknow, 1856–1877*. Princeton, NJ: Princeton University Press, 1984.

Olwig, Kenneth. *Landscape, Nature and the Body Politic: From Britain's Renaissance to America's New World*. Madison, WI: University of Wisconsin Press, 2002.

Ong, Aihwa. *Neoliberalism as Exception: Mutations in Citizenship and Sovereignty*. Durham, NC: Duke University Press, 2006.

Pagden, Anthony. *Lords of All the World: Ideologies of Empire in Spain, Britain and France c. 1500–c. 1800*. New Haven, CT: Yale University Press, 1998.

Pal, Radhabinod, and Balai Lal Pal. *The Law of Income Tax in British India: Being Act XI of 1922 as Amended by Act VII of 1939, with Explanatory Notes and Commentaries*. Calcutta: Eastern Law House, 1940.

Pal, Sayanti. "Religions Patronage in the Land Grant Charters of Early Bengal (Fifth–Thirteenth Century)." *Indian Historical Review* 41, no. 2 (2014): 185–205.

Pande, Ishita. *Medicine, Race and Liberalism in British Bengal: Symptoms of Empire*. New York, NY: Routledge, 2012.

Phillimore, R. H. *Historical Records of the Survey of the Eighteenth Century*, vol. 1. Dehradun: Survey of India, 1954.

Pinney, Christopher. *Camera Indica: The Social Life of Indian Photographs*. Chicago, IL: University of Chicago Press, 1998.

Pocock, John G. A. *The Machiavellian Moment: Florentine Political Thought and Atlantic Republican Tradition*. Princeton, NJ: Princeton University Press, 1975.

Polanyi, Karl. *The Great Transformation: The Political and Economic Origins of Our Time*. Boston, MA: Beacon Press, 1957.

Pollock, Sheldon, ed. *Forms of Knowledge in Early Modern Asia: Explorations in the Intellectual History of India and Tibet, 1500–1800*. Durham, NC: Duke University Press, 2011.

Pomeranz, Ken. "The Great Himalayan Watershed: Agrarian Crisis, Mega-Dams and the Environment." *New Left Review* 58 (2009): 5–39.

Poovey, Mary. *History of the Modern Fact: Problems of Knowledge in the Sciences of Wealth and Society*. Chicago, IL: University of Chicago Press, 1998.

The Financial System in Nineteenth-Century Britain. Oxford: Oxford University Press, 2002.

Genres of Credit Economy: Mediating Value in Eighteenth and Nineteenth-Century Britain. Chicago, IL: University of Chicago Press, 2008.

Porter, Roy. *London: A Social History*. Cambridge, MA: Harvard University Press, 1998.

Possel, Gregory L. *Ancient Cities of the Indus*. New Delhi: Vikas Publishing House, 1979.
Pottage, Alain. "The Materiality of What?" *Journal of Law and Society* 39, no. 1 (2012): 167–83.
Povinelli, Elizabeth. *The Cunning of Recognition: Indigenous Alterities and the Making of Australian Multiculturalism*. Durham, NC: Duke University Press, 2002.
Prakash, Gyan. *Bonded Histories: Genealogies of Labor and Servitude in Colonial India*. Cambridge: Cambridge University Press, 1990.
"Science 'Gone Native' in Colonial Native." *Representation* 40 (1992): 153–78.
"Subaltern Studies as Postcolonial Criticism." *American Historical Review* 99, no. 5 (1994): 1475–90.
"The Urban Turn." In *The Cities of Everyday Life: Sarai Reader* 2, ed. Ravi Vasudevan, Ravi Sundaram, Jeebesh Bagchi, Monica Narula, Shuddabrata Sengupta and Geert Lovink. Delhi: Sarai, 2002.
Prakash, Gyan, and Kevin Michael Kruse, eds. *The Spaces of the Modern City: Imaginaries, Politics, and Everyday Life*. Princeton, NJ: Princeton University Press, 2008.
Prakash, Om. "The European Trading Companies and the Merchants of Bengal, 1650–1725." *Indian Economic and Social History Review* 1, no. 3 (January–March 1964): 37–63.
Precious Metals and Commerce: The Dutch East India Company and Indian Ocean Trade. Cambridge: Cambridge University Press, 1994.
European Commercial Enterprise in Pre-Colonial India, vol. 2. Cambridge: Cambridge University Press, 1998.
Prakash, Vikramaditya. *Le Corbuiser's Chandigarh: Struggle for Modernity in Postcolonial India*. Seattle, WA: University of Washington Press, 2002.
Pramanik, Sankar Kumar. *Sundarabana: Jala, Jangala, Jibana*. Calcutta: Paribeshaka, Pustaka Bipani, 2008.
Prasad, Ritika. *Tracks of Change: Railways and Everyday Life in Colonial India*. Cambridge: Cambridge University Press, 2015.
Price, Pamela G. *Kingship and Political Practice in Colonial India*. Cambridge: Cambridge University Press, 1996.
Pritchard, Sara B. *Confluence: The Nature of Technology and the Remaking of the Rhône*. Cambridge, MA: Harvard University Press, 2011.
Rabinow, Paul. *French Modern: Norms and Forms of the Social Environment*. Chicago, IL: University of Chicago Press, 1995.
Racine, Jean, ed. *Calcutta 1981: The City, Its Crisis and the Debate on Urban Planning*. Delhi: Manohar Publication, 1990.
Rademacher, Anne. *Reigning the River: Urban Ecologies and Political Transformation in Kathmandu*. Durham, NC: Duke University Press, 2011.
Raffles, Hugh. *In Amazonia: A Natural History*. Princeton, NJ: Princeton University Press, 2002.
Raj, Kapil. "Circulation and Emergence of Modern Mapping: Great Britain and Early Colonial India, 1764–1820." In *Society and Circulation: Mobile People*

and Itinerant Cultures 1750–1950, ed. Claude Markowitz, 23–54. Delhi: Permanent Black, 2003.

"The Historical Anatomy of a Contact Zone: Calcutta in the 18th Century." *Indian Economic and Social History Review* 48, no. 1 (2011): 55–82.

Raman, Bhavani. *Document Raj: Writing and Scribes in Early Colonial South India*. Chicago, IL: University of Chicago Press, 2012.

"Sovereignty, Property and Colonization: The East India Company and Land Development in Madras." *Journal of Economic and Social History of the Orient*. Special Issue on "Repossessing Property in South Asia: Land, Rights and Law across Modern/Early Modern Divide." Forthcoming 2018.

Rangarajan, Mahesh. *Fencing the Forest: Conservation and Ecological Change in India's Central Provinces 1860–1914*. New Delhi: Oxford University Press, 1996.

Rao, Nikhil. "Uncertain Ground: The 'Ownership Flat' and Urban Property in Twentieth Century Bombay." *South Asian History and Culture* 3, no. 1 (2012): 1–25.

Ray, A. K. "A Short History of Calcutta." In *Census of India 1901*, vol. 7. Calcutta: Bengal Secretariat Press, 1902.

Ray, Amita. *Villages, Towns and Secular Buildings in Ancient India*. Calcutta: Firma KLM, 1964.

Ray, Himanshu Prabha. "The Archaeology of Bengal: Trading Networks, Cultural Identities." *Journal of Economic and Social History of the Orient* 49, no. 1 (2006): 68–95.

Ray, Mohit. *Five Thousand Mirrors: The Water Bodies of Kolkata*. Jadavpur: Jadavpur University Press, 2015.

Ray, Nisith Ranjan. *City of Job Charnock*. Calcutta: Victoria Memorial, 1979.

Ray, Rajat. *Urban Roots of Indian Nationalism: Pressure Groups and Conflicts of Interests in Calcutta City Politics 1875–1939*. New Delhi: Vikas Publishing House, 1979.

"Colonial Penetration and the Initial Resistance: The Mughal Ruling Class, the English East India Company and the Struggle for Bengal 1756–1800." *Indian Historical Review*, 2 nos. 1–2 (July 1985–January 1986): 1–105.

"The Bazaar: Changing Structural Characteristics of the Indigenous Section of the Indian Economy before and after the Great Depression." *Indian Economic and Social History Review* 25, no. 3 (1988): 263–318.

"Asian Capital in the Age of European Domination: The Rise of the Bazaar, 1880–1914." *Modern Asian Studies* 29, no. 3 (1995): 449–554.

Ray, Ratnalekha. *Change in Bengal Agrarian Society 1760–1850*. Delhi: Manohar Publishers, 1979.

Ray, Sukumar. *Folk-Music of Eastern India with Special Reference to Bengal*. Kolkata: Naya Prokash, 1998.

Reidy, Michael S. *Tides of History: Ocean Science and Her Majesty's Navy*. Chicago, IL: University of Chicago Press, 2008.

Reith, Gerda. *The Age of Chance: Gambling in Western Culture*. New York, NY: Routledge, 1999.

Rigby, Nigel, ed. *The Worlds of East India Company*. Woodbridge: Boydell Press, 2002.

Riles, Annelise. *Documents: Artifacts of Modern Knowledge*. Ann Arbor, MI: University of Michigan Press, 2006.

Robb, Peter. "Mr Upjohn's Debt: Money and Friendship in Early Colonial Calcutta." *Modern Asian Studies* 47, no. 4 (2013): 1185–271.

Robbins, Lionel, *Robert Torrens and the Evolution of Classical Economics*. London: Macmillan, 1958.

Rodgers, Daniel T. *Atlantic Crossings: Social Politics in a Progressive Age*. Cambridge, MA: Harvard University Press, 1998.

Rose, Carol. *Property and Persuasion: Essays on the History, Theory and Rhetoric of Ownership*. Boulder, CO: Westview Press, 1994.

Roswadowski, Helen M. *Fathoming the Ocean: The Discovery and Exploration of the Deep Sea*. Cambridge, MA: Belknap, 2005.

Roux, Simone, and Jo Ann McNamara. *Paris in the Middle Ages*. Philadelphia, PA: University of Pennsylvania Press, 2011.

Roy, Ananya. *City Requiem: Calcutta: Gender and the Politics of Poverty*. Minneapolis, MN: University of Minnesota Press, 2002.

Roy, Biren. *Calcutta, 1481–1981: Marshes to Metropolis*. Calcutta: National Council of Education, 1982.

Roy, Tirthankar. "Economic History and Modern India: Redefining the Link." *Journal of Economic Perspectives* 16, no. 3 (2002): 109–30.

"Rethinking the Origins of British India: State Formation and Military-Fiscal Undertakings in an Eighteenth Century World Region." *Modern Asian Studies* 47, no. 4 (2013): 1125–56.

Saberwal, Satish, ed. *Processes and Institutions in Urban India: Sociological Studies*. New Delhi: Vikas Publication, 1978.

Samuel, Raphael. *Village Life and Labour*. London: Routledge, 1975.

Sanyal, Kalyan. *Rethinking Capitalist Development: Primitive Accumulation, Governmentality and Post-Colonial Capitalism*. New Delhi: Routledge, 2007.

Sarkar, Nikhil. *Battala*. Calcutta: Dey's Publication, 1997.

Kalikata. Calcutta: Ananda Publishers, 2006.

Sarkar, Swagato. "Beyond Dispossession: The Politics of Commodification of Land under Speculative Conditions." *Comparative Studies of South Asia, Africa and the Middle East* 35, no. 3 (2015): 438–50.

Sartori, Andrew. *Liberalism in Empire: An Alternative History*. Berkeley, CA: University of California Press, 2014.

Sarwar, Mahabub, and Collin D. Woodroffe. "Rates of Shoreline Change along the Coast of Bangladesh." *Journal of Coastal Conservation* 17, no. 3 (2013): 515–26.

Sassen, Saskia. *The Global City: New York, London, Tokyo*. Princeton, NJ: Princeton University Press, 1991.

Global Networks, Linked Cities. New York, NY: Routledge, 2002.

Sayer, Daniel O. *A Desolate Place for a Defiant People: The Archaeology of Maroons, Indigenous Americans, and Enslaved Laborers in the Great Dismal Swamp*. Gainesville, FL: University Press of Florida, 2014.

Schulman, David, Sanjay Subrahmanyam and Velcheru Narayana Rao. *Textures of Time: Writing History in South India 1600–1800*. Ranikhet: Permanent Black, 2011.

Schwartz, Herman M., and Leonard Seabrooke. "Varieties of Residential Capitalism in the International Political Economy: Old Welfare States and the New Politics of Housing." In *The Politics of Housing Booms and Busts*, ed. Leonard Seabrooke and Herman Schwartz, 1–27. London: Palgrave, 2009.

Scott, Joan. *The Fantasy of Feminist History*. Durham, NC: Duke University Press, 2011.

Searle, Llerena Guiu. *Landscapes of Accumulation: Real Estate and the Neoliberal Imagination in Contemporary India*. Chicago, IL: University of Chicago Press, 2016.

Seed, Particia. *Ceremonies of Possessions in Europe's Conquest of the New World 1492–1640*. Cambridge: Cambridge University Press, 1995.

Sen, Bijayram. *Tirthamangal*. Kolkata: Parashpathar Prakashan, 2009 [1790].

Sen, Dinesh Chandra. *Studies in the Geography of Ancient and Medieval India*. Varanasi: Motilal Banarsidass, 1971.

Sen, Sudipta. *Empire of Free Trade: The East India Company and the Making of the Colonial Marketplace*. Philadelphia, PA: University of Pennsylvania Press, 1998.

Sen, Sukumar. *Battalar Chhapa O Chhabi* Calcutta: Ananda Publication, 1984.

Sengupta, Subodh Chandra, and Anjali Bose, eds. *Sansad Bangali Charitabhidhan*. Calcutta: Bangiya Shahitya Parisad, 1976.

Seth, Mesroub Jacob. *Armenians in India from the Earliest Times to the Present*. Calcutta: Luzac and Co., 1937.

Shastri, Shibnath. *Ramtanu Lahiri o Tatkalin Bangasamaj*. Kolkata: Dey's Publication, 1976.

Shell, Marc. "The Issue of Representation." In *The New Economic Criticism: Studies at the Intersection of Literature and Economics*, ed. Martha Woodmansee and Mark Osteen, 53–74. London: Routledge, 1999.

Skaria, Ajay. *Hybrid Histories: Forest, Frontiers and Wildness in Western India*. Delhi: Oxford University Press, 1998.

Simmel, Georg. *Philosophy of Money*. London: Routledge, 1978.

Simone, Abdoumaliq. *For the City yet to Come: Changing African Life in Four Cities*. Durham, NC: Duke University Press, 2004.

Sinclair, Sir John. "Address to the Reader." In *Statistical Account of Scotland*, vol. 1, ed. Donald J. Withrington and Ian R. Grant. Edinburgh: EP Publishing, 1983.

Sinha, Pradip. *Calcutta in Urban History*. Calcutta: Firma KLM, 1978.

The Urban Experience, Calcutta: Essays in Honour of Professor Nisith R. Ray. Calcutta: Riddhi-India, 1987.

Sinha, Surajit. *Cultural Profile of Calcutta*. Calcutta: Indian Anthropological Society, 1972.

Sitte, Camillo. *Der Städtebau nach seinen künsterlichen Grundsätzen*. Berlin: Springer Verlag, 2002 [1909].

Sivaramakrishnan, Kalyanakrishnan. "Ethics of Nature in Indian Environmental History." *Modern Asian Studies* 49, no. 4 (2015): 1261–1310.

Skinner, Daniel, and Leonard Feldman. "Eminent Domain and the Rhetorical Construction of Sovereign Necessity." *Law, Culture, and the Humanities* 10, no. 3 (2015): 393–413.

Smail, Daniel Lord. *Imaginary Cartography: Possession and Identity in Late-Medieval Marseille*. Ithaca, NY: Cornell University Press, 2000.
Smith, Neil. *Uneven Development: Nature, Capital and the Production of Space*. New York, NY: Verso, 2007.
Smith, P. J. "Planning as Environmental Improvement: Slum Clearance in Victorian Edinburgh." In *The Rise of Modern Urban Planning*, ed. A. Suttcliffe. London: Mansell, 1980, 99–134.
Smith, Pamela H., Amy R. W. Meyers and Harold J. Cook. *Ways of Making and Knowing: The Material Culture of Empirical Knowledge*. Ann Arbor, MI: University of Michigan Press, 2014.
Smith, Richard Saumarez. "Rule-by-Records and Rule-by-Reports: Complementary Aspects of the British Imperial Rule of Law." *Contributions to Indian Sociology* 19, no. 1 (1985): 153–76.
Soja, Edward. *Postmodern Geographies: The Reassertion of Space in Critical Social Theory*. London: Verso Press, 1989.
Thirdspace: Journeys to Los Angeles and Other Real-and-Imagined Places. Oxford: Blackwell, 1996.
Spivak, Gayatri Chakravorty. "Deconstructing Historiography." In *Selected Subaltern Studies*, ed. Ranajit Guha and Gayatri Chakravorty Spivak, 3–32. New Delhi: Oxford University Press, 1988.
"Harlem." *Social Text* 81, no. 22 (Winter 2004): 113–39.
Spodek, Howard. *Urban-Rural Integration in Regional Development: A Case Study of Saurashtra, India 1800–1960*. Chicago, IL: University of Chicago Press, 1976.
"Studying the History of Urbanization in India." *Journal of Urban History* 6, no. 3 (1980): 251–94.
Spodek, Howard, and Doris Meth Srinivasan, eds. *Urban Form and Meaning in South Asia: The Shaping of Cities from Prehistoric to Precolonial Times*. Washington, DC: National Gallery of Art, 1993.
Sreenivas, Mytheli. "Conjugality and Capital: Gender, Family, and Property under Colonial Law in India." *Journal of Asia Studies* 63, no. 4 (2004): 937–60.
Stäheli, Urs. *Spectacular Speculation: Thrills, the Economy and Popular Discourse*. Stanford, CA: Stanford University Press, 2013.
Stein, Burton. "Circulation and the Historical Geography of Tamil Country." *Journal of Asian Studies* 37, no. 1 (1977): 7–26.
Steinberg, Theodore. *Slide Mountain, or, the Folly of Owning Nature*. Berkeley, CA: University of California Press, 1995.
Stern, Philip J. *The Company-State: Corporate Sovereignty and the Early Modern Foundations of the British Empire in India*. Oxford: Oxford University Press, 2011.
"The English East India Company and the Modern Corporation: Legacies, Lessons, and Limitations." *Seattle University Law Review* 39, no. 2 (2016): 423–44.
Stewart, Tony K. *Shaping Bengali Worlds, Public and Private*. East Lansing, MI: Michigan State University Press, 1989.
Stoll, Steven. *Larding the Lean Earth: Soil and Society in Nineteenth-Century America*. New Haven, CT: Yale University Press, 2002.

Sturman, Rachel. *The Government of Social Life in Colonial India: Liberalism, Religious Law and Women's Rights*. Cambridge: Cambridge University Press, 2012.

Subramanian, Lakshmi. "Banias and the British: Role of Indigenous Credit in the Process of Imperial Expansion in Western India in the Second Half of the Eighteenth Century." *Modern Asian Studies* 21, no. 3 (1987): 473–510.

Subrahmanyam, Sanjay, and C. A. Bayly. "Portfolio Capitalists and the Political Economy of Early Modern India." *Indian Economic & Social History Review* 25, no. 4 (December 1, 1988): 401–24.

Sur, Atul. *Kalikata'r Chalchitra*. Calcutta: Shahityalok, 1982.

Taylor, Charles. *Modern Social Imaginaries*. Durham, NC: Duke University Press, 2003.

Thapar, Romila. *Shakuntala: Texts, Readings, Histories*. Delhi: Anthem Press, 2002.

Thomas, Keith. *Man and the Natural World: Changing Attitudes in England, 1500–1800*. Oxford: Oxford University Press, 1996.

Thompson, Edward P. *Whigs and Hunters: The Origin of the Black Act*. London: Allen Lane, 1975.

Timberg, Thomas. *The Marwaris: From Traders to Industrialists*. New Delhi: Vikas, 1978.

Tomlinson, B. R. *The Economy of Modern India: From 1860 to the Twenty-First Century*. Cambridge: Cambridge University Press, 2013.

Travers, Robert. "'The Real Value of Lands': The Nawabs, the British and Land Tax in Eighteenth-Century Bengal." *Modern Asian Studies* 38, no. 3 (2004): 517–58.

Ideology and Empire in Eighteenth Century India: The British in Bengal Cambridge: Cambridge University Press, 2007.

Tresch, John. "Cosmologies Materialized: History of Science and History of Ideas." In *Rethinking Modern European Intellectual History*, ed. Darrin M. McMahon and Samuel Moyn, 152–72. New York, NY: Oxford University Press, 2014.

Tripathi, Amalesh. *Trade and Finance in Bengal Presidency 1793–1833*. Calcutta: Oxford University Press, 1979.

Tsing, Anna Lowenhaupt. *Friction: An Ethnography of Global Connection*. Princeton, NJ: Princeton University Press, 2004.

Tuck, Richard. *Natural Rights Theories: Their Origin and Development*. Cambridge: Cambridge University Press, 1979.

Tyrwhitt, Jacqueline. *Patrick Geddes in India*. London: Lund Humphries, 1947.

Tyson, Geoffrey. *The Bengal Chamber of Commerce and Industry: 1853–1953, A Centenary Survey*. Calcutta: Bengal Chamber of Commerce and Industry and D. A. Lakin, 1953.

Van Antwerp, William C. *The Stock Exchange from Within*. New York, NY: Doubleday Page & Co., 1914.

Verdery, Katherine. *The Vanishing Hectare: Property and Value in Postsocialist Transylvania*. Cornell, NY: Cornell University Press, 2003.

Verdery, Katherine, and Caroline Humphrey, eds. *Property in Question: Value Transformation in the Global Economy*. New York, NY: Berg Publishers, 2004.

Verlest, Harry. *View of the Rise, Progress and Present State of the English Government in Bengal, Including a Reply to the Misrepresentation of Mr. Bolts and Other Writers*. Cornhill: J. Nourse, 1772.
Vismann, Cornelia. *Files: Law and Media*. Trans. Geoffrey Winthrop-Young. Stanford, CA: Stanford University Press, 2008.
Wahab, Amar. *Colonial Inventions: Landscape, Power and Representation in Nineteenth-Century Trinidad*. Newcastle: Cambridge Scholars, 2010.
Wallerstein, Immanuel. *The Modern World System: Capitalist Agriculture and the Origins of the European World Economy in the Sixteenth Century*. Berkeley, CA: University of California Press, 2001
Ward, Kerry. *Networks of Empire: Forced Migration in the Dutch East India Company*. Cambridge: Cambridge University Press, 2009.
Washbrook, David. "Law, State and Agrarian Society in Colonial India." *Modern Asian Studies* 15, no. 3 (1981): 649–721.
Watson, Ian Bruce. "Fortification and the 'Idea' of Force in Early English East India Company Relations with India." *Past and Present* 88 (1980): 70–87.
Weber, Max. "Stock and Commodity Exchanges." Trans. Steven Lestition. *Theory and Society* 29, no. 3 (2000 [1894]): 305–38.
"Commerce on the Stock and Commodity Exchanges." Trans. Steven Lestition. *Theory and Society* 29, no. 3 (2000 [1894]): 339–71.
"Börsenwesen: Die Vorschläge der Börsenenquetekommission." In *Max Weber Gesamtausgabe*, 1 vol. 5.2, ed. Horst Baier, M. Rainer Lepius, Wolfgang J. Mommsen, Wolfgang Schluchter and Johannes Winckelmann. Tübingen: Mohr, 1895, 558–90.
Webster, Anthony. *The Richest East India Merchant: The Life and Business of John Palmer of Calcutta, 1767–1836*. Woodbridge: Boydell Press, 2003.
White, Richard. *The Organic Machine*. New York, NY: Hill and Wang, 1995.
Whitehead, Judith. "John Locke, Accumulation by Dispossession and the Governance of Colonial India." *Journal of Contemporary Asia* 42, no. 1 (2012): 1–21.
Williams, Raymond. *Keywords*. Oxford: Oxford University Press, 1985.
Wilson, Jon E. *The Domination of Strangers: Modern Governance in Eastern India, 1780–1835*. London: Palgrave Macmillan, 2008.
Wright, Gwendolyn. *The Politics of Design in French Colonial Urbanism*. Chicago, IL: University of Chicago Press, 1991.
Yang, Anand. *Bazaar India: Markets Society and the Colonial State in Gangetic Bihar*. Berkeley, CA: University of California Press, 1998.
Yates, Alexis. *Selling Paris: Property and Commercial Culture in the Fin-de-Siècle Capital*. Cambridge, MA: Harvard University Press, 2015.
Young, Arthur. *General Report on Enclosures*. London: B. McMillan, 1808.
Zarobell, John. *Empire of Landscape: Space and Ideology in French Colonial Algeria*. University Park, PA: Pennsylvania State University Press, 2010.

Index